THIRD WORLD COUNTRIES
AND DEVELOPMENT OPTIONS: ZAMBIA

Third World Countries and Development Options: ZAMBIA

JONATHAN H. CHILESHE

VIKAS PUBLISHING HOUSE PVT LTD

VIKAS PUBLISHING HOUSE PVT LTD
Regd. Office: 5 Ansari Road, New Delhi 110002
H.O: Vikas House, 20/4 Industrial Area, Sahibabad 201010
Distt. Ghaziabad, U.P. (India)

COPYRIGHT © Jonathan H. Chileshe, 1986

All rights reserved. No part of this publication may be reproduced in any form without the prior written permission of the Author.

Printed at NEW PRINTINDIA PVT LTD, Sahibabad, U.P. (India)

To my wife Salome (Sally) and Our Children:
Chanda Harvey Jonathan
Mibela Arnold Jonathan
Nsengelwa Mutale Jonathan
Sampa Gwyneth
Chinga Gertrude

Acknowledgements

This work bears my name only in as much as a luggage tag identifies a particular item. Honesty demands that I acknowledge the contribution of all my collaborators. Perhaps I should begin with my tutelage as a citizen of the world and particularly in Zambia. Additionally, the wider exposure gained by associating with various international organizations. For instance, my spell on the Board of Directors of the African Development Bank (ADB) representing Malawi, Zaire and Zambia, the United Nations Economic Commission for Africa (ECA), the Society for International Development (S.I.D), the Association of African Public Administration and Management (AAPAM). Additionally, for the USAID grant in 1978 enabling me to do graduate work in International Financial Management at the University of Southern California and to embark on my doctoral work.

Other equally important landmarks include *inter alia* my keynote address on "National Planning of the Zambian Economy and Regional Development in East Africa" in 1971 to the Lusaka United Economics Club; on "Trade Development and the African Local Government Institutions" during the 1973 Annual Congress of the Zambia Local Government Association; on "The Role of Commerce in a Developing Society: The African Experience" during the 1981 Annual Congress of Associated Chambers of Commerce of Zimbabwe (ACCOZ): and a chapter on "Zambia" incorporated in the book *Indigenization of African Economics* by the African Association for Public Administration and Management in collaboration with Hutchinson University Library for Africa in 1981.

To be able to complete this book, far more people than I can remember, gave me encouragement and material support. And to each, I am eternally grateful. However, I crave the understanding of my cronies, in case this analysis falls short of their expectations of me. Nevertheless, it was a privilege to pit my views against my age-old friend Elias Andrew Kashita. Similarly, the inspiration of my mentor Alick (Alexander B. Chikwanda) and my eldest son Chanda.

I would be less than honest if I failed to acknowledge the initial guidance of Associate Prof. Sambhu Basu of the Graduate School of Business Administration, University of Southern California; Prof. Tony Killick and Dr. Manundu Mutsembi both of the Department of Economics at University of Nairobi, for reviewing and suggesting amends to the original manuscript; to Dr. Jack Kitson of Century University (USA) for supervising my economics doctoral programme and the dissertation "An analysis of Problems Facing Third World Countries in the Choice of Development Options in the Post-Indepedence Era".

On a wider plane, I recognize Valentine S. Musakanya, Goodwin H. Mutale, Samuel M. Molotsi, Michael C. Sata M.P., Maurice Attala, Patrick J. Chisanga, John M. Mwanakatwe M.P., Erling K. Nypan, Dr. Nicholas C. Otieno, Dr. Faith M. Kabi, Henry Barlow, A.N. Chimuka, Philip J. Box, Kayondo J. Sendi, Emmanuel G. Kasonde, Dr. Suleiman I. Kiggundu, Ato Eshetu Afework for the maps, ECA Library colleagues, Mufalo Liswaniso for generosity with editorial materials, and my editor Charles Heffernan.

I also owe much gratitude to Ms. Irene Nkembe, Woz. Yeshewatsehai Merine and Woz. Awetash Makonnen, for the preliminary typing of the manuscript at considerable personal sacrifice to themselves. However, my special gratitude and indebtedness goes to Woz. Tsehay Desta, not only for the final typing but because without her sacrifice, this task would have been a total nightmare.

I cannot thank enough through these limited lines my wife and our youngest son Nsengelwa. Both suffered greater deprivation without bitterness but with good cheer in the course of my doctoral research and in order for me to complete this work. Needless to say, the views and opinions expressed in no way represent the United Nations or the Government of the Republic of Zambia.

JONATHAN H. CHILESHE

Abbreviations

AAC	Anglo-American Corporation
AAPAM	Association of African Public Administration and Management
ANC	African National Congress
ARMB	Agriculture Rural Marketing Board
BAT	British American Tobacco
BSA	British South Africa Company
COZ	Credit Organization of Zambia
CARS	Central African Road Services
DCs	Developed Countries
DBD	Dairy Produce Board
DBZ	Development Board of Zambia
ECA	Economic Commission for Africa (United Nations)
FAO	Food and Agriculture Organization
FINDECO	State Finance and Development Corporation
FNDP	First National Development Plan
FRN	Federation of Rhodesia and Nysaland
GCDS	Grand Commander of Distinguished Service
gmb	Grain Marketing Board
GRZ	Government of the Republic of Zambia
Hon.	Honourable
IBRD	International Bank for Reconstruction and Development
IDZ	Intensive Development Zone
IFC	Industrial Finance Company
INDECO	Industrial Development Corporation
ISIC	International System of Information Classification
LDCs	Less Developing Countries
MCC	Member of the Central Committee
MLC	Member of the Legislative Council
MEMACO	Metal Marketing Corporation of Zambia
MINDECO	Mining Development Corporation
MNCs	Multi-National Corporations (also see TNCs)
MP	Member of Parliament
NAMBOARD	National Agriculture Marketing Board of Zambia

NCCM	Nchanga Consolidated Copper Mines
NIEC	National Import and Export Corporation
NHC	National Hotels Corporation
NIEO	New International Economic Order
NRG	Northern Rhodesia Government
NTC	National Transport Corporation
OAU	Organization of African Unity
PIEO	Present International Economic Order
RCM	Roan Consolidated Mines
RDC	Rural Development Corporation
RST	Roan Selection Trust
SNDP	Second National Development Plan
TANZAMA	Tanzania-Zambia Oil Pipeline
TAZARA	Tanzania-Zambia Railway
TNCs	Transnational Corporations (also see MNCs)
UBZ	United Bus Company of Zambia
UDI	Unilateral Declaration of Independence
USAID	United States Agency for International Development
UN	United Nations
UNECA	See ECA
UNO	United Nations Organization
UNDP	United Nations Development Programme
UNESCO	United Nations Educational, Scientific and Cultural Organization
UNIP	United National Independence Party
WENELA	Witwaterstand Native Labour Association
ZA	Zambia Airways
ZANA	Zambia News Agency
ZAMCAB	Zambia Taxi Service
ZECCO	Zambia Engineering and Construction Company
ZESCO	Zambia Electricity Supply Corporation
ZIMCO	Zambia Industrial and Mining Corporation
ZR	Zambia Railways
ZRT	Zambezi River Transport
ZSIC	Zambia State Insurance Corporation
ZNWC	Zambia National Wholesale Corporation
ZNBS	Zambia National Building Society
ZTRS	Zambia-Tanzania Road Services

Contents

INTRODUCTION	1
1. A CASE FOR ECONOMIC DEVELOPMENT	14
A Geo-Political Profile of Zambia	14
Mineral Resources and the History of Mining	18
The State of Agriculture	26
The Population Equation	28
The Colonial Experience	33
Pre-Central African Federation	33
The Yoke of the Central African Federation	35
The Bottom Line	37
2. LAND-LOCKED	40
The Political Origins of Zambia Becoming Land-Locked	40
The UDI's Effects on the Zambian Economy	42
Transit Routes after the UDI	46
3. DEVELOPMENT PLANNING IN ZAMBIA	51
A General Overview of Development Plans	51
History and Trends of Development Planning	59
Results	65
Conclusions	70
4. INTEGRATING THE NATIONAL ECONOMY	79
Pre-Independence Economic Imbalance	81
Pre-Independence Indigenous Business	85
Integrating Political Independence with Development	89
5. THE MULUNGUSHI AND MATERO REFORMS	98
The Mulungushi	100
Before Mulungushi	100
Mulungushi and the Industrial Development Corporation	102
Mulungushi and Small-Scale Financing	105
Mulungushi and Large-Scale Financing	107
Matero	109
Matero and the Goose with the Copper Egg	109
Directing the Mining Industry	110
Mineral Marketing	114

The Reforms and Indigenization	116
The Reforms Revisited	120
6. PARASTATALS	124
The Conceptual Framework of Public Corporations	125
Legal and Sectoral Analysis of Zambian Parastatals	127
The Mining Sector	128
The Commercial and Industrial Sector	131
The Money and Finance Sector	134
The Transport and Communications Sector	136
The Agriculture Sector	137
State Capitalism in Transition	139
The Legal Inadequacy of Using Parastatals for Development	141
7. ROADS TO DEVELOPMENT	145
Resources	147
Direct Government Sectoral Leverage	152
The Issue of Land in Development	160
Subsidies	162
Sharing Development Responsibility	165
Foreign Investment	170
CONCLUSION	176
Development Theory and Practice	176
Public Enterprises and Future Development	180
Rehabilitation	185
A Periscopic Perspective	189
APPENDICES	199
Summary of Zambian Minerals: Uses and Location	201
Summary of Zambia's Transit Routes	209
BIBLIOGRAPHY	213
INDEX	217

Introduction

To better understand many of the contemporary problems of the Third World countries, especially those constituting the developing African region like Zambia, it is essential to draw attention to certain historical, political and geographical features. These features particularly colonialism have, to a very large extent, marked these countries as unique on the world stage. Some of the effects of colonialism which require examining are the inadequacies of institutional structures, industrialization and energy resources development. Detailed analysis of each of these elements will be part of the central theme in this book. Taken together, they form a sum total of the challenges for which answers must be found in the context of selecting development options.

In examining colonialism, the book attempts to explain some of its effects on development countries' economic future, in particular the problems created by colonial heritage and the present world economic order. Many of the problems facing the economies of developing countries, including Zambia, emanate from ecological factors and from colonially imposed geo-political structures.[1] These factors made Zambia a landlocked country and created the associated problems with transit routes. Africa is unique when compared to the rest of the Third World countries when considering the features that have shaped Africa's past, present and future and its responses in the field of options in economic development.

First, the colonial powers arbitrarily partitioned the African continent without regard to African traditions, history or culture. They paid little attention to the economic and political viability of the States created. Historians and scholars of economic history have been hard put to find any rational basis for the partition of Africa. As a result, Africa has the distinction of being a huge continent with a greater number of minuscule states. Of the 48 of 50 independent African States in 1983; 7 had populations of less than 1 million; 20 had populations of between 1 and 5 million; 10 had populations of between 5 and 10 million; 9 had populations of between 10 and 30 million; and only 2 (The Federal Republic of

Nigeria and the Arab Republic of Egypt) had populations of more than 30 million.

A second unique characteristic of the African heritage was the mode of colonial penetration. In contrast to the Asian experience, colonialism in Africa was either preceded or followed by white settlement. The settlement of large numbers of Europeans from the metropolitan countries (Britain, France, Portugal, Belgium, Germany, Italy) gave them privileged positions in Africa.[2] The colonial powers enabled their settler communities to dispossess Africans of their land. This was followed by turning the native populations into farm and mine labourers of the colonial settlers. In order to ensure total and perpetual subjugation of the Africans, the colonial powers denied the former, education needed to acquire the skills necessary for autonomous economic development. The histories of pre-independence periods of such African countries as Algeria, Angola, Congo, Kenya, Mozambique, Tanzania, Zaire, Zambia and Zimbabwe help one to appreciate this point. The cumulative adverse effects of the colonial system have greatly affected the choice of development options in the post-independence era of the majority of African countries. For example, at independence in 1964, Zambia, which provided the developed world with tremendous amounts of her natural resources, had only a few hundred of its citizens with secondary school education, less than one hundred were graduates.[3] Added to the above, Zambia also inherited a weak and meagre administrative and social infrastructure.

Another important feature of colonialism in Africa was the imbalance of the economic structures it created. Most of the infrastructure set up for transport and communications especially the roads and railways networks were designed for the easy export of cash crops and the importation of manufactured goods from the metropolitan countries.[4] The production pattern was also narrowly confined to exploitation by the trading companies headquartered in the metropolitan countries like Amato Freres in Zaire (former Belgian Congo) and Unilever in most of Central and East Africa. These companies tended to concentrate in the area of plantations and mining.[5] Little or no thought was given to the need to create backward and forward economic linkages with other sectors of the economy. It is economic foundations of this kind, which independent African countries have had to correct, but without

Introduction

adequate means at their disposal. The choice of what development options to follow therefore becomes of crucial importance.

A result of these factors caused many developing countries, before and in the first years after the attainment of political independence, to pursue development plans which were dualistic in nature. Economic analysts like Samir Amin and Sir Arthur Lewis have described this dualism in the African countries as being represented by "a few industries, highly capitalized, such as mining or electric power, side by side with the most primitive techniques; a few high class shops, surrounded by masses of old styled trades; a few capitalized plantations, surrounded by a sea of peasants ... the same contrasts outside economic life. There are one or two modern towns, with the finest architecture, water supplies, communications and into which people drift from other towns and villages which might almost belong to another planet. There is even the same contrast between people; between the few highly westernized trousered natives, educated in western universities, speaking western languages, and glorifying in Beethoven, Mills, Marx or Einstein and the great mass of their countrymen who live in quite other worlds."[6]

Because African economic structures were fitted to the mainstream of the former metropolitan countries, the task of economic reorientation in the post-independence era is an uphill battle. The foundations of development of the region had been derailed or entirely neglected. Matters would not have been that bad had the bulk of the revenues not been unashamedly repatriated to the metropolitan countries. Today, the former metropolitan capitals and other cities forget that they were built, to a large extent, from plundered resources through colonialism. Yet today, they are the ones fighting hard to keep out students and migrant labourers from countries whose economies are unable to provide for them because of their colonial heritage.

Political independence in all the Third World countries, except for a handful, has been attained through strenuous effort. However, socio-disengagement from the old economic order has not been at all effective. Developing Africa, unlike the other developing regions of Asia and Latin America, is increasingly dominated by the developed countries, especially through the latter's transnational and multinational corporations. Africa's dependence on international trade and the export of primary commodities is relatively higher than that of other developing regions. This affects development options as the region becomes increasingly dependent on foreign trade,[7] capital, technology and expertise.

There is enough evidence not only of the importance of foreign trade in the economy of a country but also of the existing correlation with development trends. It is quite evident that the Third World countries have not been able to fully utilize their foreign trade potential. Trade during the colonial era in many Third World countries was "marginalized" in terms of setting up of global development objectives and priorities. It was this marginalization which gave birth to the "centre-periphery" concept or the "unequal exchange vision" which was strengthened even in the aftermath of independence between the Third World countries and the industrialized countries. In other words, the Third World countries were geared towards developing an absorptive capacity for manufactures from industrialized countries which in turn supply, mainly, primary raw materials of agriculture and minerals.

Study after study stress this fact. An investigation in eleven countries by Michael Michaely has demonstrated that there is a direct relationship between export expansion and economic growth. For instance, between the average size of annual changes in the ratio of exports to GNP with the average annual change in per capita GNP.[8]

The economic weakness of the Third World countries who depend on external trade, as well as foreign financial and technical assistance for their economic development, has contributed to the emergence of strong linkages between the former metropolitan powers and their former colonies. Thus, the strong degree of interdependence existing between the European Economic Community (EEC) and Africa, especially as members of the Africa, Caribbean and Pacific Group (ACP). These relations were given a new dimension with the creation of the Treaty of Rome in 1958, which made provisions under the Yaounde Conventions. The Lome Convention of 1979[9] was a marked improvement in comparison to the Yaounde Conventions because the former represented an understanding by the EEC of the need to help the ACP States develop more rapidly, with special emphasis on industrial development and trade. However, it should not be overlooked that there is very little in Lome's terms to ensure that the

Introduction

ACP states become totally autonomous economies in the long term.

The nature and character of the externality of economies of the Third World countries exemplified by the above is rooted in the legacy of colonialism. It underpins as the accompanying chart shows, why Third World countries are most likely to continue to apply development theories nurtured outside their own respective regions. The degree of externality is perpetuated both by traditional and non-traditional axis as shown in the chart.

The former exists between the North (which is a sum total of OECD countries of the expanded West with the confined East) and the South (members of G.77 and the Non-Aligned Movement). The non-traditional axis in global dimensions is that forged or being forged between countries of the South (within the context of the Group of 77[10] and the Non-Aligned Movement). Similarly, by the former with the Eastern Block, whether as members of the Council for Mutual Economic Assistance (CMEA) or with China. Consequently, association in any form or with any of the various groupings is an option which has to be made taking account of some likely implications, on each country's rate of economic development and the benefits to be derived therefrom. Merits and demerits of association briefly referred to above is a sum total and can either assist or retard endogenous development. However, this depends on the development option selected by each country and the ability to implement a chosen path of economic development.

The imperatives of the present world economic order in the context of interdependence, have been presumed in certain quarters, to be better than during the colonial era for a large number of the Third World countries. However, there is ample evidence in this analysis to show to the contrary. In other words, this is not so much because of resort to development theories. For many Third World countries, especially producers and exporters of primary agriculture and mineral products, the world economic scene has not altered that much. Hence, the formalization and non-formalization of economic arrangements as exist through membership of the United Nations, continued North-South dialogue can not be said to have improved the economic fate of many Third World economies to any significant degree. As a group, Third World countries continue to be suppliers of primary raw materials to the industrialized countries and importers of manufactures and services from the same sources. This has been perpetuated by the hegemony of the developed industrialized countries over the developing countries. The end result is that the majority of the Third World countries (except for the New Industrializing Developing countries like India, Brazil, Mexico, Singapore) have become increasingly dependent rather than interdependent with the developed countries.[11]

The above facts illustrate the difficulty for policy-makers in choosing developing options in the African region. It is paradoxical that a continent as richly endowed with an abundant variety of resources, like Africa, should harbour nearly 40 to 50 per cent of the world's poorest least developed developing countries (LDCs). The African region is among the world's richest in mineral and hydro-power resources. It has 97 per cent of the world's known chromium resources, 96 per cent of its diamonds, about 50 per cent of the gold, 48 per cent of the cobalt, and about 41 per cent of all bauxite reserves. Africa's iron-ore reserves are great and are estimated to be about 10 per cent of the world total. The continent's agricultural land is about 190 million hectares, without the inclusion of forests and other non-arable lands.

However, wealth of the kind referred to above is lost to African countries, in part, due to the effect of colonialism.

Currently, Third World countries are economically incapacitated by the developed countries. The majority of developing countries' currencies are inconvertible and are linked to currencies of the former metropolitan countries. Their international transactions, even among themselves, are inevitably via the currencies of the developed countries. The forced linking of the currencies of developing countries to those of the developed countries has created a sort of an umbilical connection, which has its roots in colonialism. Changes in the value of the currency of a developed country automatically affects the currency value of developing countries, especially in terms of international purchasing power. A good example is provided by the existing links between the French currency and that of the former French African colonies. The Communaute Financiere Africaine (CFA) currency, administered by the Banque Centrale des Etats de l'Ouest (BCEAO) in former French West Africa (except Mali), and Banque des Etats de L'Afrique Centrale (BEAC) in former French Central Africa, is closely linked with the French

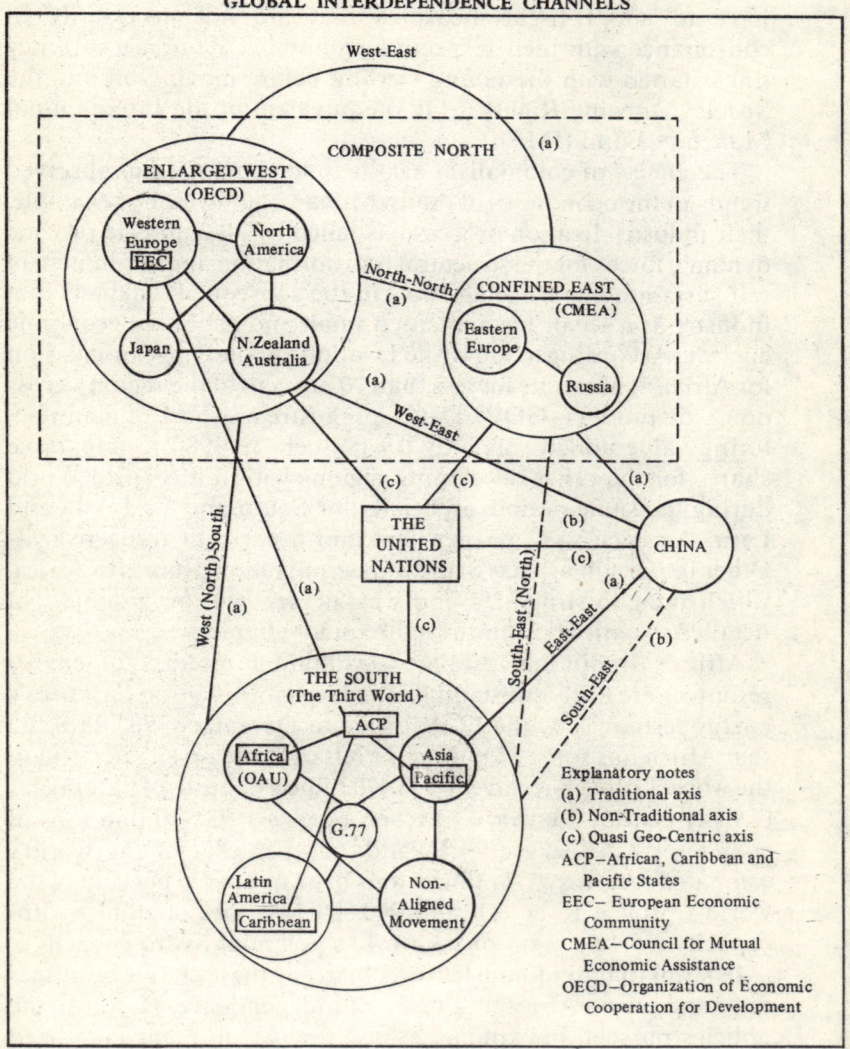

franc. The French treasury has underwritten the CFA and so it is easily covertible into the French franc. CFA countries inevitably have to adopt fiscal measures that are not necessarily in consonance with their respective economies. Zambia's currency links started with the pound sterling before moving on into the Special Drawing Rights (SDRs) equivalent of the International Monetary Fund (IMF).

The effects of colonialism also help to explain other observed trends in the economies of many African countries. For example, their industrialization process has failed by all counts to provide dynamic forces for the structural transformation and attainment of self-sustainment. We shall show in the subsequent chapters that industry as a sector has remained small and at best an economic enclave. According to the United Nations Economic Commission for Africa, it accounts for less than 10 per cent of the region's gross domestic product (GDP). Developing Africa's share of manufacturing value added was only 0.9 per cent in 1980. Comparable shares for the other developing regions within the Third World during the same period, especially for South and East Asia and Latin America were 2.7 per cent and 6 per cent respectively.[12] Other important aspects of the industrialization process in Africa, which include structures and weaknesses, will be a subject of detailed examination in the subsequent chapters.

Africa's traditional and non-traditional reserves of energy resources are fairly substantial. The latest information on Africa's energy resources by the ECA, issued in December 1982, indicate that Africa has some 55 billion barrels of crude oil (8.5 per cent of the world's proven reserves); 208,470 billion cubic ft. of natural gas (7.9 per cent of the world's proven reserves); 88.5 billion tons of coal reserves* (between 1.16 and 3.05 per cent of the world's estimated reserves); 1.7 million tons of uranium (25 per cent of the world's proven reserves); and 200,000 MW of potential hydro-capacity (35.4 per cent of the world's potential hydro-capacity).

In terms of the colonial legacy, obviously these energy resources did not grow on Africa as a result of independence. However, the policies pursued in exploiting these reserves in many colonized countries should be scrutinized. Actions by metropolitan enterprises like the British South Africa (BSA) Company, con-

---

* For coal, South Africa accounts for about 81 per cent of Africa's reserves.

centrating their production energies in exploiting the coalfield at Wankie in Zimbabwe (former Southern Rhodesia) to total indifference to exploiting similar coal deposits in Northern Rhodesia raise considerable suspicion about metropolitan motivations. In other words the archives of colonialism must continue to be searched in order to establish the missing logical links.

Discussion will be devoted to closely examining the African development crisis as created by or contributed to by the external and internal powers outside their control. The crisis of development and especially the choice of development options is complicated by the sense of urgency to fulfill the hopes of the newly independent societies.

In studying Zambia in particular, her abundance of mineral wealth, especially her copper, lead, zinc and cobalt, was a major stumbling block to the early attainment of political independence. The wealth explains the stronghold on the economy by expatriates and settlers, arguments against the dissolution of the Central African Federation of Rhodesia and Nyasaland (FRN), and the behind-the-scene suggestions to cede part of Northern Rhodesia to Southern Rhodesia.

Zambia has recently risen to prominence as a world copper producer. Minerals other than copper have yet to be explored and evaluated in terms of their respective quantity and use as contributors to the country's economic development. Zambia's experience will help to show some of the dangers, advantages and disadvantages of over-reliance on a few resources. The analysis will help the reader appreciate another side of the importance of natural resources. This is, that Zambia could well have existed as a political entity even without its copper resources.

To a large extent, Zambia's economic structural deformities at independence were a result of defective colonial policies. Economic development was structured in a way which favoured the production of primary agriculture products and extractive industries for export to the metropolitan power and its associates. Concentration on the production of a few primary commodities left Zambia economically vulnerable to external forces.

Challenges facing the economy in the early years of political independence included: attempting to liberate the economy from continued foreign domination (caused by the long colonial era and entrenched during the time of the Federation of Rhodesia and Nyasaland (1953-1963); finding ways of minimizing overdependence on copper, which generated the greatest proportion of domestic activity in terms of employment, income, foreign exchange earnings for development; lessening undue reliance on protecting the economy against industrialized countries manipulating the country's economic growth and development; containing the adverse effects of being land-locked which was made worse by Rhodesia's UDI barely a year after Zambia's attainment of independence; and not least, choosing an appropriate ideology or philosophy.

Zambia's early years of independence will probably go down in history as its most difficult period. A crucial event of both a political and economic nature was the Unilateral Declaration of Independence by Rhodesia in November, 1965. Perhaps the only positive aspect of this action was that it brought home to the Zambian body politic the realties of its situation. The event became instrumental in starting a process of internal and external economic reforms. For instance, construction of the Tazama Oil Pipeline and the Kafue and Kariba North Bank Hydro Power Stations were initiated, several internal and international highways were improved, the Tanzania-Zambia Railway (TAZARA) was begun, the Maamba Coal Mines were opened and construction if the Mwembeshi Earth Satellite Station was also begun.

The above partial list of problems and needed actions show some of Zambia's challenges and dilemmas in the post-independence era. Zambia's politicians and technocrats, like their counterparts in other developing countries, were responsible for reorienting the economy in the context of overall economic development. They had to realise that their task required great resolve and dedication. The country's development plans will be viewed from the point of view of their objectives, such factors as: increasing the Gross National Product (GNP); maintaining of price stability (in order to control the adverse effects of galloping inflation and keep it from lowering the purchasing power, especially for the lower income group of the community); striking an equitable balance in terms of provincial development; reaching or full employment and the full utilization of national resources; and attaining an equilibrium in external trade.

Zambia's various development plans compare favourably with others. Planners attempted to address themselves to the same issues facing other developing countries. To a large extent, planning was considered as an attempt at providing opportunities for a better life for all sections of the community and the removal of glaring inequalities. An important step was deciding which economic theory to espose for development effort.

The analysis will focus on the contribution of the State and its battery of public or parastatal organizations. The analysis includes a review of the role played by private enterprises in contributing to Zambia's economic destiny. Among the big private enterprises are the activities of the Anglo-American Corporation (AAC), the Roan Selection Trust (RST), the British South African (BSA) Company and certain expatriate commercial banking institutions. There will also be an assessment of the contribution of individuals and indigenous private enterprises.

Research for the thesis included, apart from analysing the references and bibliographies appended therein, extensive interviews with key economic and political figures. Brain-storming sessions were held with colleagues working on similar issues, with professors and with my doctoral programme supervisor.

Chapter 1 is devoted to the resources on which Zambia's economic development was fashioned. It is a macro-economic survey of both human and other resources. Chapter 2 briefly analyzes some implications of the country's geographical location. Attention is therefore focused on Zambia being a landlocked country. The analysis is set against as historical pattern, to show some of the adverse effects on the economy. The UDI has particular significance here.

Chapter 3 begins with a brief survey of patterns and background to the theories of development plans in the plans of various Third World Countries. It then goes into an in-depth analysis of some of Zambia's various development plans in the ten years of independence. It reviews how Zambia tries to build economic confidence signposts. Attention is focused on the appropriateness of using foreign economic development models, like a kitchen recipe in a developing country.

Chapter 4 attempts to link, in a progressive manner, several consecutive periods in the country's economic reform process. The analysis attempts to provide a prelude to subsequent chapters through an explanation of number of governmental economic decisions.

Chapter 5 contains a detailed analysis of the issues raised in the previous chapters. It concentrates on how the policy-makers responded and attempted to correct mistakes or imbalances of the past in their attempts to maintain economic growth and development. The analysis also reviews some of the causes for several contradictions, and how much use was made of existing and/or created institutional machineries.

Chapter 6 focuses attention on the role played by public or state corporations as tools for fostering economic development. Events are contrasted with experiences in neighbouring countries. Several economic sectors are reviewed in this same vein. There is also an analysis of the country's own legal structure as a catalyst in the development process.

Chapter 7 deals with the issue of development options in the Third World countries against the background of the Zambian experience. It attempts to show what Zambia has been faced with in restructuring an economy based on one primary resource. Another problem touched upon is the role of foreign developmental inputs. The chapter analyzes results of attempts used in advancing the rate of economic development in the first decade of political independence in the African region.

The book concludes with an overview of Zambia's experiences during a decade of indepedence. It illustrates some of the tasks which face developing countries' aspirations for genuine development. The experiences of other developing countries which attained independence later that Zambia is also compared. Mauritius for example, tended to place greater emphasis on *travail pour tous* (work for all). These countries have had the advantage of witnessing lessons being learned by their predecessors. Seeking to initiate formal planning at district and village or ward levels is the belief by many developing countries that it is in those areas where the problems are most urgently felt.[13] Tanzania, independent before Zambia, has had time to try more experiments in pursuing development. Conclusions about the Zambian experience can apply to a great many other developing countries.

REFERENCES

1. J.D. Fage, *An Atlas of African History*, Edward Arnold (Publishers) Ltd., Great Britain, 1958; J.D. Hargreaves, *Prelude to Partition of West Africa*, Macmillan, London, 1970; W.E.F. Ward, *Emergent Africa*, George Allen and Unwin Ltd.,

Introduction
13

London, 1971; Roland Oliver and J.D. Fage, *A Short History of Africa*; Penguin African Library, 1963.
2. Walter Rodney, *How Europe Underdeveloped Africa*, Bogle L'Ouverture Publication, London, 1972, p. 174.
3. Jonathan H. Chileshe, "Zambia", *Indigenization of African Economies*, Adebayo Adedeji (ed.), Hutchinson University Library for Africa, London, 1981, pp. 87-132.
4. The analysis stresses the correlation between exports and economic growth by demonstrating the effect of growth in exports as a reliable engine of growth in those economies where growth of export exhibits a high overall economic growth. This can not be the case for most developing countries where growth of exports is determined by market forces which the latter is not capable of influencing significantly.
5. Richard I. Sklar, *Corporate Power in African State*, University of California Press, 1975.
6. Sir Arthur W. Lewis, "Economic Development with Unlimited Supplies of Labour", *Manchester School of Economics*, May 1954 quoted in A. Hirchman in his *Strategy of Economic Development*, Yale U.P. 1958, p. 126.
7. Bela Belassa, "Exports and Economic Growth", *Journal of Development Economics*, Vol. 5, No. 2, June 1978, pp. 181-189.
8. Michael Michaely, "Exports and Growth: An Empirical Investigation", *Journal of Development Economics*, vol., 4, pp. 49-56.
9. Under article 131 of the Treaty of Rome, African countries who had been former colonies of France, Belgium and Italy and had attained political independence were allowed to continue in a privileged position in their economic relations with the European Economic Community (EEC). A later development of this lack of delinking between some 10 industrialized countries of Europe and 64 Former Colonies was the Conclusion on 31 October 1979 of the Lome II Conventions and 1983 between 12 and 66 respectively under Lome III.
10. The Group of 77 (G.77) started as an informal group of seventy-seven member developing countries in 1964. Its membership by 1983 had increased to about 126 countries. By the time of going to print, the G.77 has evolved a structure with nuclei in Geneva (UNCTAD), Vienna (UNIDO), Paris (UNESCO), Rome (FAO) and at the United Nations in New York. The G.77 serves as the principal organ of the Third World in articulating and promoting the Group's collective economic and political aspirations and interests.
11. Leaster B. Pearson, *Partners in Development*, Praeger Publishers, New York, 1969, p. 3.
12. Economic Commission for Africa, *ECA and Africa's Development 1983-2008: A preliminary perspective study*, Addis Ababa, April 1983, p. 12.
13. Jonathan H. Chileshe, "An Extension of the Development Paradigm: Grassroots Experiments and People's Movements in Africa", *18th World Conference*, Society for International Development, Rome, 1-4 July 1985.

Chapter One

A Case for Economic Development

The attainment of our political freedom in 1964 equipped us for the aggressive pursuit of economic freedom, for to have one without the other is to be a one-legged nation in a two-legged world.[1]

The general impression provided by this observation is quite unmistakable. In other words, there is more to attainment of political independence than the merriment of celebrations, when the old flag is lowered and the new one is hoisted in its place. This is why we must begin by identifying the economic basis for development for any developing country. The next possible logical step is to address the scope of possible options which might have been available to the policy-makers in similar positions. Consequently, the analysis will enquire into the merits and demerits of adapting or rejecting foreign economic development models.

Zambian experience in this field is useful to other developing countries who may be trying to re-orient their economies, especially where they use similar techniques. Consequently, we will look at the following: resource mobilization and deployment; geographical location; external trade relations, especially reliance on a single export commodity; the indispensability of transit routes; the emergence of a national philosophy, and the general restructuring of the economy to obtain both sectoral and provincial balances. These problems create the background for studying the challenges and dilemmas faced in the aftermath of independence.

A GEOPOLITICAL PROFILE OF ZAMBIA

Zambia's geopolitical boundaries are a result of the behind-the-scene activities of the former colonial powers. The position and the landlocked status of Zambia and the attendant economic impli-

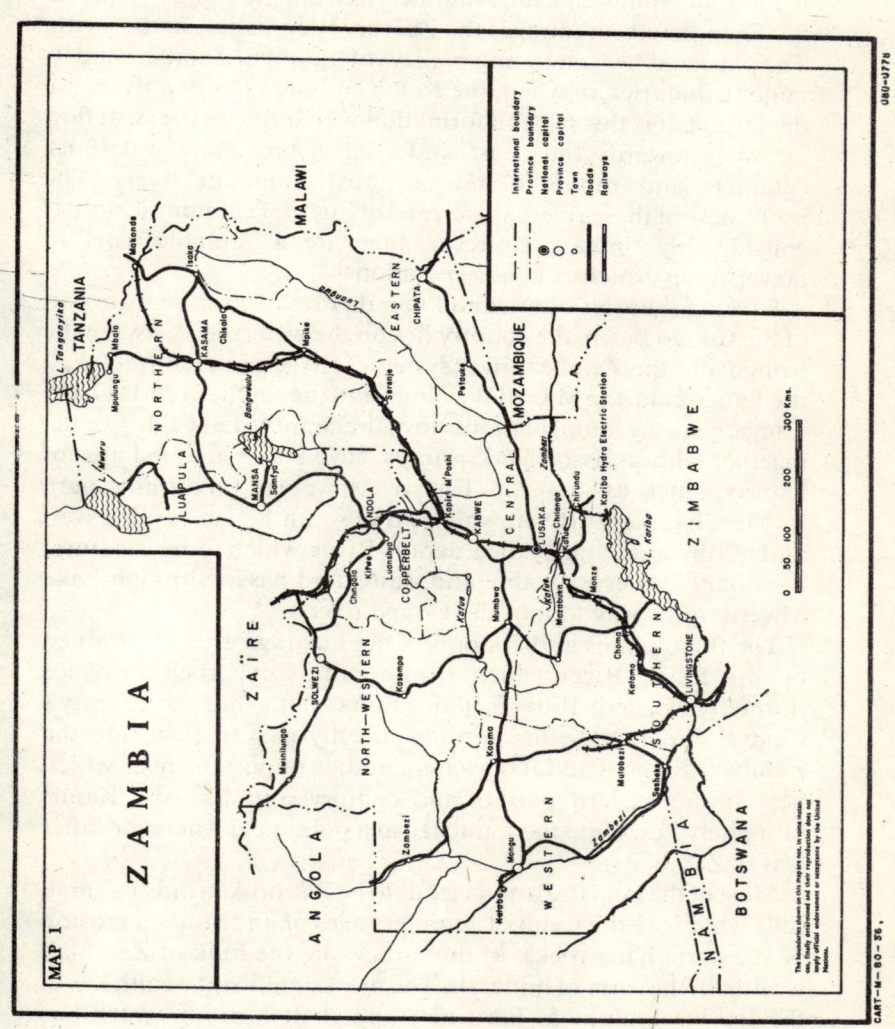

cations, cannot be divorced from the interaction of history and the manoeuvres of colonial administrators.

Zambia's landscape falls southwards from the Congo-Zambezi divide in the north towards the Zambezi depression in the south. The plateau is broken by huge valleys of the upper Zambezi and its major tributaries, of which the Kafue and the Luangwa rivers are the largest. It is this plateau formation which causes the swift flow of water towards the coast and creates the many waterfalls, cataracts and rapids found on most Zambian rivers. The usefulness of these rivers as a means of internal communication is considerably limited. However, they are a valuable asset in powering hydro-electric power stations.

Except for the Northern and Luapula Provinces, which are part of the Congo Basin, the country lies on the watershed between the Congo and the Zambezi river systems. Three great natural lakes, the Bangweulu, the Mweru Wa Ntipa and the southern end of Lake Tanganyika are found in the north of the country. Lake Bangweulu, together with its associated swamps, covers an estimated area of approximately 6,080 sq. km. The area is drained for the most part, by the Chambeshi River, which flows south before turning west and northwards, and by the Luapula River which forms a natural boundary between Zambia and Zaire, and passes through Lake Mweru on its way to join the Congo river.

The flood plains and swamps of the Lukanga area are drained by the Kafue River which rises in the Copperbelt Province (formerly Western Province) and flows south into the country's Central Province before turning directly east to flow into the Zambezi River. The Luangwa is another important river which lies in the eastern part of the country and like the Kafue, ultimately flows into the Zambezi, at a point near where the latter leaves Zambia.

Most of the land in Zambia tends to be flat, broken only by small hills which are the result of countless ages of undisturbed erosion of the underlying rocks. In these rocks lie the bulk of Zambia's wealth, in the form of minerals. The most significant wealth lies in the 160 km long by 50 km wide copperbelt corridor, which has continued to be the mainstay of the Zambian economy. There are some parts of the country where the range and the quantity of mineral resources has not yet been established. For instance, the Western Province, (formerly Barotse Province) which to a very large extent is covered by the sands of the **Kalahari Desert**.

The state of being landlocked is analysed in greater detail in a subsequent chapter since, it is one of the main factors that influenced the country's development in both political and economic terms during the first decade of independence.

Also to be looked at are questions of land, population, capital inflow and the creation or restructuring of certain institutions. These are the bases for Zambia's economic growth and development. Also are external forces which have affected the economy. The consequences of the unilateral declaration of independence (UDI) in neighbouring Southern Rhodesia were particularly great. All these elements provide a basis against which to measure the challenges and dilemmas recounted in this book.

Economic indicators place the Zambian economy in the group of "middle-income economies", with a per capita GNP in 1980 of US$560 and an annual growth rate of 0.2 per cent between 1960 and 1980. The average annual rate of inflation between 1960 and 1970 was 7.6 per cent and between 1970 and 1980 was 8.1 per cent. Adult literacy in 1977 had climbed to 44 per cent from 39.4 per cent in 1969 and 28.5 per cent in 1963. These indicators show that Zambia, like other developing countries of the Third World was attempting to achieve certain basic economic goals. Those goals were a growth of GNP, price stability, an increase in adult literacy and an increase in food self-sufficiency as well as full productive employment.

History has yet to produce a magic formula to which a country can turn for immediate solution to its problems. To move from being a low-income country to high-income country requires tremendous efforts on many fronts. Even exceptional circumstances do not preclude great effort. Libya's sudden discovery of oil gave that economy an unprecedented economic growth, but not much development had taken place. Sufficient time has to elapse before one state of affairs gives way to another. The duration of such a transition is in most cases a function of many elements and on how well they can be combined to bring about the desired result. Zambian policy makers realised that development takes a long time. They knew it was not only a question of how long a country has been politically independent. There is a lot more to economic development than the mere attainment of political independence and the availability of natural resources, especially in the absence of the right mix of the human endeavours.

There can be no denying that political independence is an important event in shaping the course of economic events in formerly colonised states. This act opens up options needed for tackling the issues affecting economic independence. Independence gives a government the ability to formulate and execute its own social and economic policies, whether the models used are indigenous or not. What is really important is that the government have full control of the country's direction and in setting its objectives. In other words, a truly independent country must determine its own pattern of economic development.

MINERAL RESOURCES AND THE HISTORY OF MINING

The history of Zambia's resources is the history of the country's important minerals. Oral and written accounts of some of the early explorers have added substance to some of the stories handed down from generation to generation. Each account attests to the fact that minerals were an important item of trade in the central African region. In 1591, Flipo Pigafelta reported that he had found ample evidence of this in the Kingdom of the Congo and that trade in minerals had been going on for a very long time in the area.

Flipo Pigafelta's account was based on information of a Portuguese traveller, Odoardo Lopez and on its translation by Hakluyt which refers to an area east of the Congo River—including what is now Zambia: "In this Kingdome, the Anziques, there are many Mines of Copper ...the men are active and nimble and leape up and down the mountains like goats..." Reference is also made to "the mines of Bembe" which could possibly be associated with the old name for Lake Bangweula and with the Bemba people.[2]

By the end of the 1880s, the existence of minerals in the interior of Africa was made known to the outside world by explorers, missionaries and traders. The accounts of missionaries like David Livingstone lend a lot of truth to the fact that copper articles were being traded from Central Africa as far north as Uganda and by the Arab slavers to the west and east coasts of Africa by the middle of the 19th century. A few copper ornaments have been found in several prehistoric burial places. Some of these date back to the beginning of the 15th century. Research conducted on behalf of the former Rhodes-Livingstone Institute (now a research arm of the University of Zambia) provides further evidence of the long history of trade in minerals in the region. The discovery by archaeologists of copper ornaments as part of the burial pieces at Ingombe Ilende[3] in present Zambia adds to these facts. The importance of such a revelation is the distance of the burial site from the copperbelt region and the proof it provides of the importance of mining over the history of the region.

Systematic prospecting for minerals by the British South Africa (BSA) Company is relatively recent in comparison with similar activities elsewhere in the world, and was not actively pursued until the company received the Royal Charter in 1889. Similar prospecting by the then Northern Rhodesia Government (NRG) was not initiated until 1950, when a grant was made by the Colonial Development and Welfare Fund to the then Northern Rhodesia Department of Labour and Mines. This led to the setting up of a separate Geological Survey Department. Since its inception, the Geological Survey Department was able to carry out major projects in mapping out the country's mineral wealth. The two principal mining companies, the Anglo-American Corporations (AAC) and the Roan Selection Trust (RST), also made extensive explorations of the country's mineral wealth. Only rarely was this important activity undertaken in joint cooperation between the government and the mining companies and then only in collaboration with outside bodies like the United Nations Development Programme (UNDP).

The granting of the Royal Charter to the Northern Territories (BSA), Exploring Company, one of the numerous subsidiaries of the Bechuanaland Exploration Company formed by Edmund Davis in association with Cecil Rhodes strengthened the interests of British merchants against other competing European commercial adventurers in the area. It was the nature of colonial diplomacy that the Royal Charter conferred an unchallengeable prerogative to its recipients. It also lent authenticity to the concessions and treaties concluded by British citizens with the native population. Research into some of these treaties, especially the ones purported to have been signed between the BSA Company and the Lozi Paramount Chief on the mineral concessions, has raised several questions. One of the conclusions drawn from that research is that the old Zambian native rulers showed either immaturity or an absence of good judgement. Another conclusion is that early British commercial adventurers tricked native leaders and adopted a deliberately dishonourable behaviour. That behaviour was contrary to civilized standards of respect for the body and spirit of an agreement.

Research has also shown that the old African rulers were conversant with their respective jurisdictions and aware of their territorial boundaries and knew the extent to which their rule could be enforced.[4] Evidence of this kind disputes the inferred interpretation of the BSA Company that the Lozi Paramount Chief could ever have sanctioned prospecting concessions or rights to the Company for areas extending so far afield as to encompass the whole of the copperbelt region, over which he had no jurisdiction. African chiefs were able to verify the area of their respective authority by established practices of law enforcement, for instance, through homage to their throne by the subject people. There is no evidence to show that the Lamba people of the region in question paid tribute or homage to any Lozi chief at the time of the treaties. Neither is there any claim of such homage in any Lozi history.

Backed with the articles of the Royal Charter, the colonial commercial adventurers found it possible to make honourable what otherwise would have been dishonourable acts. It was the desire of these colonialists to achieve certain concealed objectives. A restatement of parts of the preamble to the Charter of Incorporation of the BSA Company will clarify this line of thought. The Charter stated, among other things, that the existence of a powerful British company, controlled by British subjects, and having its particular field of operation in that region of South Africa lying to the north of Bechuanaland and west of Portuguese East Africa, would be advantageous to the commercial and other interests of the subjects of the United Kingdom.

This charter inspired the BSA Company to hold, use and retain for its own purposes, the full benefit of all concessions and agreements. However, the validity or interests of the other parties especially the local authorities, was of no consequence.

A brief list of the country's mineral deposits is at Appendix A. The most talked about Zambian mineral has been copper. The tendency therefore is to forget about such other important minerals as nickle, amethyst, cadmium emeralds, monazite,

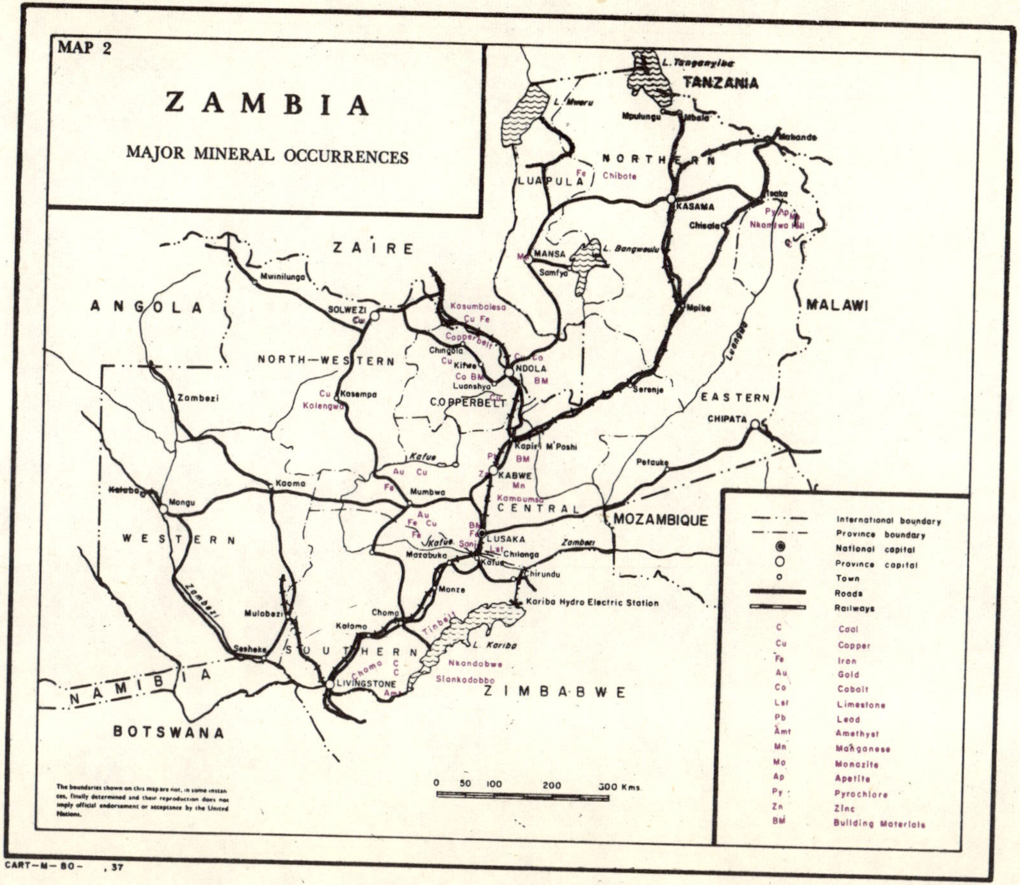
MAP 2
ZAMBIA
MAJOR MINERAL OCCURRENCES
ZAIRE
ANGOLA
TANZANIA
MALAWI
MOZAMBIQUE
ZIMBABWE
BOTSWANA
NAMIBIA
NORTHERN
LUAPULA
NORTH-WESTERN
COPPERBELT
CENTRAL
EASTERN
WESTERN
SOUTHERN
L. Tanganyika
L. Mweru
L. Bangweulu
L. Kariba
Mpulungu
Mbala
Nakonde
KASAMA
Chibote
Isoka
Chisala
Nkombwa Hill
MANSA
Samfya
Mpika
Mwinilunga
SOLWEZI
Kasumbalesa
Copperbelt
Chingola
Kitwe
Luanshya
NDOLA
Serenje
Kasempa
Kolengwa
Zambezi
Kapiri M'Poshi
KABWE
Kamaumsa
CHIPATA
Petauke
Kafue
Kaoma
Mumbwa
Mongu
Kalabo
LUSAKA
Chilanga
Mazabuka
Chirundu
Kariba Hydro Electric Station
Monze
Choma
Kalomo
Mulobezi
Sesheke
LIVINGSTONE
Tinbelt
Nkandabwe
Siankadobbo
0 50 100 200 300 Kms
International boundary
Province boundary
National capital
Province capital
Town
Roads
Railways
C Coal
Cu Copper
Fe Iron
Au Gold
Co Cobalt
Lst Limestone
Pb Lead
Amt Amethyst
Mn Manganese
Mo Monazite
Ap Apetite
Py Pyrochlore
Zn Zinc
BM Building Materials
The boundaries shown on this map are not, in some instances, finally determined and their reproduction does not imply official endorsement or acceptance by the United Nations.

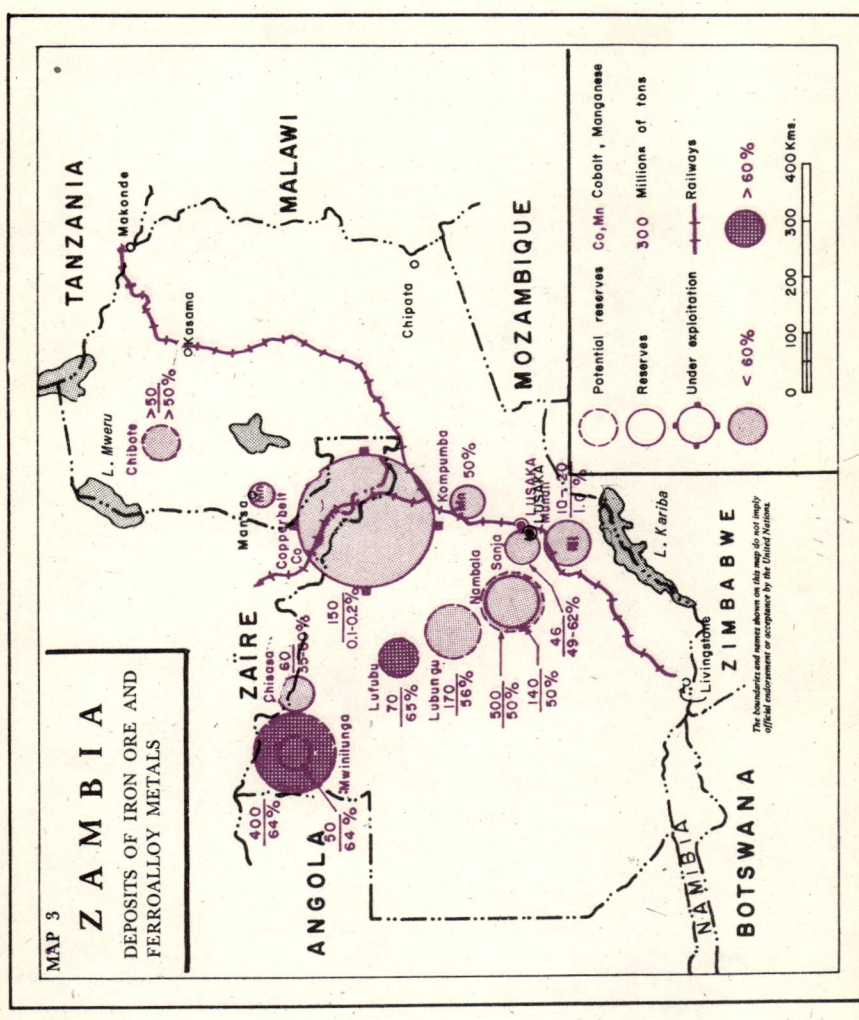

apatite, pyrochiore, and other industrial ores such as those used in cement and fertilizers. Table 1.1 reviews Africa's mineral deposits against a projected world demand for these minerals. This information can help the reader appreciate Zambia's world standing and what problems and prospects arose in the decade under review.

We can get a better picture of the future with a closer look at facts and problems pertaining to minerals. Data in Maps 2 and 3 could be taken together with data in Table 1.1 to clarify that picture.

Mining is likely to continue to be a very important economic activity in Zambia's development. Proceeds from mining, especially copper and cobalt accounted for over 90 per cent of the country's total foreign exchange revenues between 1964 and 1974. Data in Table 1.1 show world-wide copper reserves. Zambia's share is estimated at between approximately 700 and 800 million short tons of ore, with about 2.92 to 5.17 per cent purity. If the other minerals were fully exploited, the contribution of the mineral sector to the economy could have been considerably enhanced.

Deliberate and increased exploitation of the country's mineral potential could have given Zambia an advantageous economic position, especially in reducing her dependence on outside supplies of certain imported mineral products. It could also have enhanced the economic diversification. Mineral exploitation could have been linked to the country's agricultural potential, which is one of the pillars of development in any economy. Increased fertilizer consumption could benefit a large portion of the economy if it were manufactured locally, and would have resulted in considerable foreign exchange savings. Apatite, a member of the phosphate group found in Zambia, could have been used in fertilizer production.

Data in Table 1.2 underpin the importance of mining as Zambia's most important sector in the economy. Zambia has no proven oil deposits. However, exploration for oil especially natural gas continues to be undertaken in the Western Province which is an extension of the sands of the Kalahari Desert. Also other sectors of the economy need to be developed into major contributors to the Gross Domestic Product (GDP). In its first decade, the country was faced with the problem not only of maintaining production capacity, but also determining what other minerals, apart from copper, were to be developed.

TABLE 1.1: African Minerals and Forecast of World Mineral Demand*

Mineral	Africa's reserves to world (%)	Africa's production to world (%)	Forecasts of world demand (year and quantity) 1985	Forecasts of world demand (year and quantity) 2000	Specifications
Diamond	90.0	76.6	...	...	
Chromium	97.0	32.8	...	...	th. short tons
Phosphate	72.0	28.9	207,000	456,000	th. short tons
Gold	64.0	79.2	44,300	64,700	th. troy oz.
Manganese	55.0	32.2	14,800	22,500	th. short tons
Cobalt	42.0	75.5	87,500	143,400	th. pounds
Bauxite	33.0	6.0	...	...	
Uranium	28.0	20.9	106,700	193,200	short tons U
Copper	13.0	20.0	16,600	30,000	th. short tons
Tin	7.0	8.4	323	404	th. long tons
Zinc	6.0	5.1	9,620	13,300	th. short tons
Lead	3.0	6.2	7,940	11,650	th. short tons
Iron ore	3.0	9.5	750	1,130	mn. short tons

* *Source: Mineral Facts and Problems,* US Bureau of Mines 1974-1975 and *Bureau of Mines Bulletin 667:* Table of summary forecasts of US and rest-of-world demand 1973-2000.

TABLE 1.2: Copper's Relative Importance in the Economy
(Value and Percentage)
1968 — 1974

	1968	%	1970	%	1972	%	1974	%
Gross Domestic Product (GDP) (K million)	1,062	100.00	1,258	100.0	1,312	100.0	1,820	100.0
Contribution to GDP (K million)								
(i) Mining & Quarrying	412	38.8	460	36.6	324	24.7	622	34.2
(ii) Copper	411	38.7	457	36.3	318	24.2	616	33.8
(iii) (ii) as % of (i)		99.8		99.3		91.1		99.0
Government Revenue (K million)								
(i) Copper and others (total)	306	100.0	432	100.0	315	100.0	628	100.0
(ii) Copper	183	59.8	218	50.5	69	21.9	337	53.7
Export (f.o.b) spectrum (K million)								
(i) Total	518	100.0	669	100.0	538	100.0	900	100.0
(ii) Copper & Cobalt	497	95.9	634	94.7	501	93.1	847	94.1
Employment intake ('000)								
(i) Total (paid sectors)	319	100.0	343	100.0	365	100.0	386	100.0
(ii) Mining industry	55	17.2	58	16.9	58	15.9	63	16.3
(iii) Copper industry	48	15.0	48	14.0	51	14.0	56	14.5
Indigenous & as % of (iii)	43	89.6	14	91.7	46	90.2	52	92.9
Expatriates & as % of (iii)	5	10.4	4	8.3	5	9.8	4	7.1

Source: Monthly Digest of Central Statistical Office, April/May 1975. Zambia Mining Year Book, 1974. Report and Statement of Accounts for Year ended December 31st. Bank of Zambia, 1976.

Both geology and accessibility compounded these problems, especially in the initial stages. Other economic considerations had to be taken into account before deciding of when to open a particular mine. They included determining the grade of the mineral content, the size and life-span of the deposit in question. It was necessary to determine whether the mineral could be obtained as a by-product of another mineral or a metal extraction process. For instance, gold and cobalt extraction, in most Zambian mines, are by-products of copper extraction.

The mutual proximity of many of the country's known minerals, as shown in Maps 1 and 2, could lead one to misleading conclusions. Development of a mine enterprise requires enormous finances and high technical skills, both of which are scarce in a developing country. The decision to borrow and import skills implies continued reliance on foreign factors to sustain the rates of growth and development. The lack of indigenous skills and finance compelled Zambia to continue to rely on expatriate assistance in developing its mineral resources, a situation which, after independence, creates conflicts of interest between the host country and the foreign companies. This was inevitable when the latter felt ill at ease with the country's emerging political ideology. However, Zambia was incapable of dispensing with expatriate control over mining so long as such control ensured the inflow of needed foreign exchange earnings.

Such a state of affairs gives the upper hand to the foreign company, both in determining and influencing production alternatives. The fact that the major companies were multinationals meant that they could pick and choose which minerals they preferred to exploit, based on profitability in the world markets rather than on consideration of the economic development of the host country. During the ten years of the existence of the Central African Federation of Rhodesia and Nyasaland between 1953 and 1963 the two mining companies of AAC and RST neglected exploiting Zambia's coal deposits and concentrated on those at Wankie in Zimbabwe (formerly Southern Rhodesia). Prior to the creation of the Federation, which was to all intents and purposes a union. Northern Rhodesia had no balance of payments problems. However, during the Federation days, 56 per cent of net investment went to Rhodesia and only about 40 per cent to Zambia. Through this arrangement Northern Rhodesia lost an annual average of about 24 million pounds sterling in revenues derived from exploitation of her copper industry.

Resolving this issue is one of the many challenges of economic development that faced Zambia during the first decade of independence. That the country wanted to assume early control over its natural resources and increase ownership of the major business concern was no secret. However, the government also made no secret of its interest in receiving assistance from foreign companies in extending and expanding the mining activities. It was bent on earning additional tax revenues, which could be used to help some of the country's unemployed. Such honest expressions tended to scare away foreign investments. The promulgation of the liberal investment codes that followed were designed to attract foreign investment in mining. However, these codes were treated with contempt by foreign investors.

The lack of foreign investment coupled with the lack of local skills for carrying out proper feasibility studies explains some of the difficulties which arose in setting up the local Zambian iron and steel complex. Zambia has the basic ingredients for an iron and steel industry. It has enough power resources from hydro-electric power stations, like the giant Kafue and Kariba North Bank generating stations. There are sufficient coal deposits at Maaba and Kandabwe, which are accessible by the completion of the railway spur extension from Choma to Masuku. Also, the completion of the Tanzania-Zambia Railway brought the coal deposits discovered in the extreme northeast area in the Luangwa Valley into the picture. There are ample iron ore deposits, despite their rather low mineral content in certain locations.

Conclusions drawn from various feasibility studies point to the theoretical soundness of setting up of an iron and steel industry, with an expected annual capacity of 100,000 tonnes in the first stage. That steel could meet Zambia's growing internal demand. The growth of an engineering sector in the Zambian economy could also lead to a number of altered economic perspectives. For one thing, the dependence on the southern suppliers would be considerably reduced. The decision to proceed or not was another of the many challenges facing Zambian policy-makers. It was realised that an efficiently operated domestic iron and steel industry would have considerably lessened the hardship brought to bear on the economy in the wake of UDI, especially in relieving the capacity constraints on the transport network. Efficiency is a function of factors which did not exist in Zambia at the time. The project remained on the drawing board and even the initial infrastructure which had gone into constructing the electricity power lines were later dismantled at great financial loss to the economy.

Perhaps a final note on mining ought to draw on the deliberations of the Conference of Ministers of ECA held in Accra from 19 to 23 February 1973. The meeting underscored the importance of mining with the adoption of resolution 218(X) on ownership and control of natural resources. The resolution cited, *inter alia* the importance of greater explorations, exploitation, and utilization of mineral resources. It also stressed the need for establishing forward and backward national industrial links to the mining industry and the promotion and strengthening Governments' capabilities to negotiate advantageous contracts in respect to exploration and exploitation of mineral resources. Zambia had already made progress by acquiring a substantial degree of ownership and management of its mining sector by the time of the above resolution.

THE STATE OF AGRICULTURE

This analysis will be brief because agriculture is given further analysis in subsequent chapters. Agriculture constitutes for many African countries by far the largest section of economic activity. For example, in 1963, it was 24 per cent in Southern Rhodesia, about 65 per cent in Malawi and about 70 per cent in Zambia. The United Nations Economic Commission for Africa (ECA) and the Food and Agriculture Organisation of the United Nations (FAO) estimated that about two-thirds or more of the African population is engaged in agriculture and that it accounted in 1964 for 39.4 per cent of the total GDP (without including the Republic of South Africa).

A feature of African agriculture during the colonial era and prior to independence was the high proportion of subsistence production in the total agricultural output. Subsistence production in most of countries of the continent, accounted for over half of the total value of agricultural production.

A Case for Economic Development

Agricultural activities include the production of agricultural products, animal husbandry, fisheries and forestry. Zambian agricultural production, excluding animal husbandry, fisheries and forestry, can be put into three groups, namely, non-food products, food products which are partly or mainly for export, and food products mainly for the domestic market. The main products in the non-food group prior to independence were tobacco and a little cotton.

Zambia's major food products, which were produced partly for export and partly for domestic consumption, are maize and groundnuts. Other food crops like rice, cassava, millet, sweet potatoes and sorghum are for domestic consumption and highly localized. Agriculture did not enjoy much government support. Consequently, yield per acre in Zambia was comparatively lower than in Southern Rhodesia, where the administrator gave great encouragement to increased acreage and yield per acre.

The production of cash crops such as wheat and potatoes calls for irrigation. Irrigation schemes did not attract the attention of the colonial regimes. Such an omission in this crucial area in pre-independent Zambia raises many unanswered questions. Why, for instance, did planners of the time ignore this important aspect in a country that boasts of perennially flowing rivers and an abundance of lake water.

The evidence is incomplete on why agricultural production in Zambia was what it was. Available data from agencies such as FAO tend to indicate that the slow growth of agricultural output up to the mid-1960s was a general phenomenon in developing countries. It must not be forgotten that demand and supply tend to go hand in hand. Agricultural output during the pre-independence era was not completely studied and became yet another of the new nation's challenge in fostering economic growth and development.

The rate of growth of the Zambian economy like those of other developing countries producers of primary commodities is affected by agricultural output. Agriculture must provide food supplies and raw materials if industries are to expand at a fast rate. Also, labour has to be freed for industrial jobs in the mines and in the urban centres. Advanced agriculture adds to the expanded tax base needed to finance and expand the country's infrastructure (roads, railways, electricity, telecommunications, bridges, etc.) and other development projects. Increased agricultural output raises the income of the subsistence population and their purchasing power for the products of domestic industries. Last but not least, advanced agricultural output results in food self-sufficiency, reducing food imports while keeping pace with high rate of population growth.

In the light of the above, (additional evidence provided in the subsequent chapters) there is ground for concern and suspicion as to why development of Zambia's agriculture was not given more priority. This sector should have been given greater prominence in the pre-independence era as well. Had such a course been taken, agricultural development might not have been among the most formidable challenges in Zambia's post-independence era.

THE POPULATION EQUATION

The country's population structure is another problem. Classical economic theory lists among the essential ingredients to production not only land and capital, but also labour, which is based on a country's population. The labour component is an invaluable resource in achieving economic development. A country's population supplies the labour force and consumes what is produced.

The 1963 census estimated Zambia's population at about 3,490,000. The 1969 census revealed an increase in population to about 4,057,000, representing an annual growth of about 2.7 per cent. During the intercensal period, the country witnessed an unprecedented growth in urban population of around 48 per cent, estimated at an annual average rate of 9 per cent[5]. Conclusions in the 1964 UN/ECA/FAO Economic Survey Mission on the Economic Development of Zambia, attempted to forecast the country's future urban population growth rates, indicated that it was likely to rise by between 4.9 and 5.3 per cent between 1963 and 1980. In point of fact, the annual growth rate of the urban population during that period turned out to be far greater than was anticipated; it rose by approximately 15 per cent per year. A similar study by Patrick Ohadike in 1974 assumed on the basis of constant fertility projections observed that Zambia's population growth would be in the order of 4,710,000 in 1974, 5,499,000 in 1979, 6,474,000 in 1984, 7,726,000 in 1989, 9,314,000 in 1994 and 11,315,000 by 1999[6]. This population pressure was concentrated in ten towns, especially those along the railway. The main challenge for the government in this issue was finding employment for these newly urbanized people.

The above figures don't tell us whether the labour force was expanding as rapidly as the total population. Population trends are vital in a country bent on a course of economic development. Planning for education, setting up industries and the construction of roads and railways, needed to speed up the process of economic growth, must take account of population trends.

Population distribution in a large country (746,000 sq. km) doesn't give us an accurate picture of the population equation. Numbers can be meaningless where a high degree of illiteracy and a lack of skill prevails. Zambia's population is relatively small for the land available. The general pattern of population distribution is one of great dispersion and fragmentation. The rural areas account for over 70 per cent of the country's total population. The rate of migration to the urban centres increased with the advent of political independence. This created a population concentration in the few urban centres along the main railway line from the Victoria Falls Bridge to the Copperbelt.

The population equation, in terms of the manpower utilization is crucial if economic development is not to lose its direction. Among the salient features of the country's population distribution are not only its quality but also its potential effects on output and productivity, both in the short and long term. In other words, the analysis must look into the question of manpower utilization and requirements. This cannot be done on the basis of Ricardo's labour theory value*, but rather, in terms of the Keynesian outlook which stresses the governments' concern for full employment as a step towards a welfare state.

The Zambian government undertook several studies in the area of population. In 1969, the Development Division of the Office of the Vice-President issued a study covering a wide range of economic aspects. The study examined Zambia's manpower

* David Ricardo's labour theory postulates the value of a commodity as being proportional to the amount of labour expended during the production process of the final product. However, this has since been discredited with the advent of technology where the labour content in the finished product can only be computed remotely and indirectly. Labour is no longer the major influencing factor in most highly sophisticated and complicated production lines.

needs in terms of sustaining the country's major industries. It also looked into the various sectors to determine possible sources of skilled labour supply. It studied the country's educational and vocational training institutions, constraints arising out of a rural population concentration and the bases to be used to measure productivity.[7] Unfortunately, implementation of the proposed solutions was not fully dealt with.

The country's education programme stressed increased literacy, employment needs and the replacement of non-Zambians who had long dominated the economic scene. "The most immediate and perhaps the most serious handicap facing Zambia at independence", stated the Government in 1966, "had been the lack of trained manpower due directly to the unprogressive and restrictive educational policy" before independence. The following figures show the dominance of non-Africans in professional occupations in the early 1960s: architects, 98.5 per cent; civil engineers 99.6 per cent; electrical engineers, 100 per cent; mechanical engineers, 100 per cent; chemists 99.1 per cent; physicians and surgeons, 96 per cent; pharmacists, 98 per cent; lawyers, 98.2 per cent; accountants, 94 per cent. Consequently, secondary school enrolment, about 14,000 at independence, had to be increased to no less than 109,000 by 1980. Similarly, the national university was required to produce no less than 1,000 graduates per year by 1980.

The term 'labour force' appears frequently in this book. It implies all those above the age of childhood who are either working or seeking a job, especially those working for a wage or salary. In other words, labour force equals employers, the self-employed, unpaid family workers in industry, and rural subsistence workers. The labour force at independence was approximately 1.5 million, or 42.8 per cent of the population. Fewer than 300,000 out of the 1.5 million of the men and women making up the country's labour force had jobs in the industrial sector. Thus, more than 1.2 million, or 80 per cent of the labour force, were in the non-industrial, mostly rural sector.

Table 1.3 shows a high concentration of the industrial labour force in the mining industry. However, hard statistical evidence of labour problems during this analysis have been extremely limited. Various sample surveys on which data in Table 1.3 are also based provide only a blurred picture. The nation's problems include the need to resolve under-employment of the working poor in the

A Case for Economic Development

TABLE 1.3: Population Dispersion in Zambian Industries
(1973)

A. Aggregated Sectors

Sectors	'000
Manufacturing	48.1
Mining and quarrying	61.1
Electricity, gas and water	4.7
Total industries	113.8

Source: *UN Demographic Yearbook* 1976; *UN Yearbook of Industrial Statistics* 1975; *UN Statistical Yearbook* 1976.

B. Selected Sectoral Disaggregation

ISIC	Industry	'000
230	Metal ore mining	58.2
290	Other mining	2.7
	Sub Total	60.9
311/2	Food products	11.8
313/4	Beverages and tobacco	3.9
321	Textiles and spinning/weaving	2.8
322	Wearing apparel	4.8
323/4	Leather products and footwear	0.7
331	Wood products	1.4
332	Furniture etc.	1.5
341	Paper products	0.7
342	Printing and publishing	1.7
351	Industrial chemicals	1.1
352	Other chemicals	2.3
353	Petroleum refineries	0.4
355	Rubber products	1.2
356	Plastic products	0.3
361/2	Pottery/China/glass and products	0.3
369	Non-metal products	3.4
371	Iron and steel	0.8
372	Non-ferrous metals	0.1
381	Metal products	5.2
382	Machinery	1.0
384	Transport equipment and motor vehicles	1.2
390	Other industries	0.2
4101	Electricity	3.4
420	Water works and supply	1.2

Source: *UN Yearbook of Industrial Statistics.* 1975. Vols. I & II.

urban and rural areas where some members of the household are employed while others are without a source of income. Basically, a symptom of the past pattern of development or the so-called legacy of colonialism. Additionally, the need to resolve the problem of frustrated job-seekers. Based on the 1969 census provided by the various labour exchanges it was about 380,000. These figures do not include the expected inflow from secondary school leavers. This helps explain some of the policies adopted by the government towards mobilizing the nation's idle human and natural resources, especially in the rural sector. Plans were therefore set afoot by the Government to stimulate the nation's key productive sectors in order to absorb in the mainstream about 25,000 or two per cent of additional workers per year.

The country's policy makers had some room for optimism in terms of population and development. After all there were large areas of unused land available. They had identified a growing local consumption potential for domestically produced agricultural products. There was enthusiasm in government over economic reforms. A general feeling emerged that the great potential locked up in an idle labour force could be released. Given the country's abundant natural resources, the picture looked bright.

Improvement in the quality of the labour force leads to higher productivity. Changes in productivity of any labour force could mirror likely organizational changes in work or an introduction of new methods or equipment or both. Higher productivity can also come about as a result of changes in economics conditions. Generally however, productivity will tend to increase rapidly in periods of vigorous economic growth and to lag in periods of economic stagnation. The question therefore would be to establish whether it is proportionately or inversely related to population and employment.

Zambia's concern with the population equation stemmed in part from realization that a definite link exists between poverty and population. Rapid population growth which is not accompanied by available productive job opportunities contributes to poverty notwithstanding its contribution to the labour pool. What Zambia needed most was to devise some means of keeping both economic development and population growth. It attempted to achieve these two goals by formulating a fairly comprehensive set of social and economic development policies some of which are discussed in greater detail in Chapter Three.

THE COLONIAL EXPERIENCE

There is a long chain of events surrounding the Zambian colonial experience sometimes known as the scramble for Africa. The Berlin Conference helped in shaping present Zambia's political boundaries.

Apologists for the humane side of colonialism are quick in reminding the uninitiated reader of the civilizing influence of missionaries like David Livingstone, who preached universal brotherhood. What is often forgotten by such spokesmen is the fact that these words were for the consumption of the oppressed native populations, not the oppressors.

It is false reasoning to assume that the primary objective of railways and roads constructed in the aftermath of missionary exploration were intended to benefit the native populations. It is also equally untrue that roads and railways were built in tropical Africa to enable European knowledge of hygiene, medicine and dietetics to reach the so-called primitive natives of the "Dark Continent." Rather, as evidence has shown in many former colonies, the primary aim was to exploit the newly discovered raw materials needed to fuel industrial progress in the more developed world. The transport infrastructure was not built to create smooth internal communications networks within Zambia or the rest of tropical Africa.

Pre-Central African Federation

The history of Zambia's colonial period begins with classic mercantile exploitation. The commercial adventurism of men like Cecil John Rhodes and the activities of the British South Africa (BSA) Company are a part of that story. Rhodes' ambition was to build a railway from the Cape in South Africa through Central African territories to Cairo and to turn all these lands over to British sovereignty.

To this end he managed to secure the Charter incorporating the BSA Company in 1889. The company's operations were later enlarged by an extension of the Charter in 1891, to cover territories north of the Zambezi River.

The colonial administrative structure carved the territory into segments. The first region was North-Western Rhodesia. North-Western Rhodesia dates back to concessions obtained by Ware and Lochner from the Lozi Paramount Chief who is purported to have requested, in 1889, to be placed under British protection (perhaps because of the influence exerted on him by the missionary Francois Colliard). This action was followed with the appointment by the British Foreign Office, at the insistence of the BSA Company, of a British resident to Barotseland in 1897.

The second segment was the region of North-Eastern Rhodesia, which dates back to the time of the extension to the Charter in 1891. The BSA Company followed this up by purchasing concessions and signing treaties with native chiefs. A direct consequence of these activities was the appointment, in 1891, of Sir Harry Johnston as Her Majesty's Commissioner and Consul-General for British Central Africa. He assumed control over North-Eastern Rhodesia until 1895, when through another agreement, the BSA Company appointed Major P.W. Forbes as its own Deputy Administrator.

The BSA Company acquired other land concessions from the African Lakes Company (the Tanganyika Estate) and from the Carl Wiese Mineral Concession (the North Charterland Concession). The latter was apparently obtained through negotiations with Chief Mpezeni of the Ngoni in 1891. The Company failed in its attempts to acquire the area of Katanga. Others like Alfred Sharpe and Joseph Thompson also at Rhodes' behest obtained numerous concessions from chiefs between the Luangwa and Luapula rivers and as far south as present Ndola.

North-Eastern Rhodesia and North-Western Rhodesia remained distinct administrative sectors until the Northern Rhodesia Order in Council of 1911, amalgamated them as one territory as Northern Rhodesia. L.A. Wallace was appointed Administrator under the jurisdiction of the High Commissioner for South Africa. The Northern Rhodesia Order in Council, following the Devonshire Agreement of 1923, terminated the Chartered Company's administration in 1924. Thereafter, the Crown assumed administrative responsibility over Northern Rhodesia and granted it protectorate status as a territory. This was accomplished through the creation of an office of Governor, responsible for overseeing the affairs of the new protectorate.

A common feature of the colonial era was its lack of attention to aspects which would have laid a good foundation for future economic development. Zambia was no exception. Very little was done to provide an infrastructure upon which economic development could later be built. The construction of all-weather roads and the extension of the railway head from Bulawayo across the Zambezi River at the Victoria falls to Kabwe (formerly Broken Hill) were done in colonial self-interest. The same is true of the railway line constructed from Livingstone to Mulobezi and its associated Zambezi River Transport (ZRT) which was designed to facilitate the transportation of timber and logs from Mulobezi which were used as railways slippers for the maintenance and extension of the Rhodesia Railways network. According to Walter Rodney these actions were to under-develop Africa.[8]

Zambia's continued dependence on foreign factors in the aftermath of independence is a legacy of colonialism. Most glaring of the colonial shortcomings were the inadequacy of the educational system and infrastructure facilities. African education in Northern Rhodesia was far inferior to that provided to non-Africans. However, shortcomings in the struggle for economic development cannot and should not be predicated on the colonial legacy alone. The Central African Federation analyzed in the next part of this analysis, added to problems for the Zambian economy after the attainment of independence.

The Yoke of the Central African Federation

The creation of the Central African Federation or the Federation of Rhodesia and Nyasaland, in 1953, was part of the grand design of British colonialism. Cecil John Rhodes would not have been disappointed had he lived to see some of the fruits of his dream. The economic implications on the post-independence economy are difficult to quantify in absolute terms. There has been some assessment of the adverse effects brought to bear on the Zambian economy by the Federation.

The ills suffered by the Zambian economy by being forcefully integrated in the Central African Federation are too numerous to recount. It contributed to retarding economic progress in the aftermath of independence, especially in the first ten years. The cost of the Federation on Zambia was high in fianacial, political and social terms.

For instance, the federal structure hurt the Zambian economy long after it had been dismantled as a legal entity. Its ghost haunted the entire Zambian economic fabric, in particular because the future Zambia was given very low priority in the

Federation's economic infrastructure, except where it served the interests of expatriate investors in Rhodesia, South Africa and overseas. Similarly, African education and agriculture were relegated to the sidelines in terms of federal fund allocation. Preference was shown for Salisbury and Southern Rhodesia in general, instead of Lusaka and Northern Rhodesia when it came to funding economic development projects. Salisbury, not Lusaka, was chosen as the seat for the federal university. The Kariba-South Bank, rather than Kariba-North Bank or Kafue Gorge, were selected as sites for construction of the first federal hydropower station. This decision was surprising because the main argument in raising the loan with IBRD, and the main consumer of the end product, was the Zambian copper mining industry.

Southern Rhodesia derived considerable benefits from the federal structure, far in excess of her contribution to the federal coffers. The legal structure of the Federation restricted Northern Rhodesia's ability to take necessary corrective actions to protect its economy from unbridled exploitation. Northern Rhodesia was denied the means of appropriating a large part of its copper royalties as a result of the 1950 agreement providing for direct payment of 20 per cent of all royalties to the Government. Also, under the structure establishing the Federation, Northern Rhodesia could not unilaterally raise the rate of income tax on copper companies. As a way of softening the blow of exploitation, the territorial government was allowed to apply a territorial surcharge, limited to 20 per cent of the basic rate of income tax. This device was not intended to give the territory greater yield in revenues, but rather was a measure intended to hoodwink Northern Rhodesia, since the surcharge was set by the federal legislature.

Territorial governments were confined to providing for African education and agriculture, with a very restricted budget. The principal instrument for raising revenues was reserved for the federal Government. At the time of the Federation's break-up, Southern Rhodesia as the federal capital accounted for over 55 per cent of total gross fixed capital formation for the whole Federation.[9] Thus, Southern Rhodesia government did not suffer to the same extent as the other two unequal partners because it, unlike the others, derived direct and indirect benefits from most federal expenditures by virtue of hosting the federal capital and in being a battery of federal institutions.

The federal fiscal system denied the territorial economy the greater part of revenue from local copper sales and other mineral revenues notwithstanding the provisions of Article 95 of the Federation Constitution which was intended to extend direct financial assistance to Northern Rhodesia. Under Article 95, each territorial government received a certain percentage of revenue from the common federal income tax pool collected from each of the constituent territories. Zambia received an average of approximately 17 per cent per annum. This was considerably disproportionate in view of her large contribution to the common pool. This observation is supported by the conclusion reached in the Report of the Federal Fiscal Commission, undertaken pursuant to Article 95 of the Federal Constitution. That Report stated that most of the provisions of Article 95 had remained a dead letter.

It would have been possible to minimise the extent of economic injury to Zambia had it been so desired by those in power. One of the ways would have been to enforce Article 95 of the Federal Constitution as pointed out by the Fiscal Commission. However, industries in Southern Rhodesia mushroomed as a result of having a large protected export market in both Northern Rhodesia and Nyasaland. On the dark side of the scale, Northern Rhodesia suffered a net loss in revenue of approximately £100 million between 1953 and 1963.[10]

This evidence helps to show that for Zambia there was a direct link between political events and difficulties in attaining economic development in the aftermath of independence. It would have been political suicide for the African political leaders to have used the state of economic backwardness as a reason to postpone the attainment of political independence. Herein lies the wisdom of Kwame Nkrumah's political advocacy exhorting colonized countries in first Seeking Political Freedom and assuming the attainment of economic freedom as a logical sequence to independence.

However, the dilemmas and challenges in attaining economic development were compounded by independence itself. The country inherited a lopsided economic structure, had less than 100 university graduates, and entered an economic arena over which it had little control.

The Bottom Line

The preceding analysis has shown the kind of economic situation which Zambia inherited at independence. It was most unsuitable for purposes of facilitating both economic growth and development. Basically because its educational system was much below requirements as there were only a handful indigenous people with anything like adequate training to be able to take their place in government service. The transport system was linked in such a way that the Zambian economy was subservient to Southern Rhodesia and led to all kinds of inconveniences and frustrating difficulties especially in the aftermath of Unilateral Declaration of Independence by Rhodesia. Zambia's main supplies, particularly of petrol and oil came in by way of the southern routes as did coal and coke for the mining industry. Zambia was also highly dependent for electricity on the jointly owned (with Southern Rhodesia) Kariba Dam whose power station is situated on the south-bank in formerly Rhodesia. Trade, another important engine of economic growth and development for any economy was so organized that Zambia seemed irrevocably linked to supplies from Southern Africa, not only for raw materials, but also for ordinary consumer goods of every-day life.[11]

An inheritance is both a legacy of assets and liabilities. However, the preceding summary of the situation at independence shows that Zambia inherited more liabilities than assets. Many of the inherited situations from the colonial era had to be put right. It was therefore assumed that the attainment of political freedom in 1964 had equipped Zambia for the agressive pursuit of economic freedom. "For to have had one without the other," as President Kenneth Kaunda observed during the 10th Anniversary of Independence was tantamount to being "a one-legged nation in a two-legged world".[12]

To a large extent, the problems of implementation are the hardest to solve. Development problems against the above historical background were more than technical. Otherwise they could not have been immune to marginal adjustments by the end of the first decade of independence.

Certain conclusions can be drawn from the above brief analysis, especially with regard to the interaction of forces and their impact on Zambia in the aftermath of independence. These are the factors that constituted the base for some of the country's dilemmas and challenges. In addition, colonial forces had conspired to weaken the new country. Most of the treaties alleged to have been signed with Zambian Chiefs were mere deceptions in the same way, road and railway traffic was purposely designed to defraud the Zambian economy.[13]

REFERENCES

1. President K.D. Kaunda, *Zambia 1964-1974: Ten Years of Achievement*, Zambia Information Service, Lusaka, 1974, p. 1.
2. Roan Consolidated Mines Limited, *Zambia's Mining Industry: The First 50 Years*, Zambia, 1978, p. 14.
3. Brian M. Fagan (ed.), *A Short History of Zambia*, Oxford University Press, Lusaka, 1966, p. 96.
4. Yu M. Kobisbehanov, "Media for Communicating Geographical Information in Pre-colonial Africa", *II International Congress of Africanists*, Moscow, 1967, p. 4.
5. Republic of Zambia, *Second National Development Plan*, Ministry of Development Planning and National Guidance, Lusaka, December 1971, p. 9.
6. Patrick O. Ohadike, *The Population of Zambia*, 1974 World Population Year, C.I.C.E.D. Series, p. 154.
7. Development Division, Office of the Vice-President, *Zambian Manpower* Government Printer, Lusaka, 1969.
8. Walter Rodney, *How Europe Underdeveloped Africa*, Tanzania Publishing House, Dar es Salaam, 1972, pp. 223-272.
9. United Nations, *Zimbabwe Towards a New Order: An Economic and Social Survey*, United Nations, 1980, p. 19.
10. Richard Hall, *The High Price of Principles: Kaunda and the White South*, Hodder and Stoughton, London, 1969.
11. Republic of Zambia, *First National Development Plan—1966-1970*, Government Printer, Lusaka, 1966, p. v.
12. President K.D. Kaunda, *op. cit.*, p. i.
13. Richard Hall, *op.cit.*, p. 17.

Chapter Two

Land-Locked

Once the scramble had begun it became largely a matter of prestige or pride to govern colonies; and many people in Europe passed easily from the idea of governing colonies to the idea of owning them.[1]

Zambia was formerly known as Northern Rhodesia. It has an area of approximately 752,620 sq. km. or 290,586 sq. mile and therefore larger than the combined area of Belgium, France, Holland and Switzerland. It is the seventh largest territory in south central Africa and lies approximately between latitudes 8 and 18 degrees South and between longitudes 22 and 24 degrees East.

The name Zambia comes from the River Zambezi which rises in the north-west corner and also forms the country's southern boundary.

THE POLITICAL ORIGINS OF ZAMBIA BECOMING LAND-LOCKED

Zambia's connection with British rule began with the missionary travels of men like Dr. David Livingstone from 1841 until his death at Chitambo in 1873.[2] The origins of Zambia becoming land-locked was also due to British international connections. The origins of Zambia's structure of the economy on the other hand, were with the administration established by the British South Africa (BSA) Company when from about 1890 the area was divided into two separate administrative entities: North-West Rhodesia on one hand and North-East Rhodesia on the other. These were later amalgamated into Northern Rhodesia in 1911. However, direct administrative powers of Northern Rhodesia by the British Government through its Colonial Office were not assumed until 1924.

A country is by definition land-locked when it is totally dependent on access to the sea through other states. It was therefore, at Berlin Conference and British manoeuvring during the colonial era together with event connected with the Scramble for Africa which led to Zambia being land-locked long before the attainment of political independence in 1964. Zambia is land-locked and has boundaries with Zaire, mainland Tanzania, Malawi, Mozambique, Zimbabwe, Botswana, Angola and the Caprivi Strip—an extension of South-West Africa (Namibia).

A lack of free access to the sea is cause for many political, economic and social development problems. Land-locked countries must therefore cultivate a sense of good neighbourliness with transit states. A transit state is any state with or without a sea or ocean coast but situated between a landlocked country and the sea or ocean and through which traffic of the land-locked state must pass to the outside world. Consequently, any failure to cultivate good neighbourliness only adds to the difficulties and aggravates the adverse effects of the development process of land-locked countries.[3] There are 12 other land-locked countries in Africa: Botswana, Burundi, Central African Republic, Chad, Lesotho, Malawi, Mali, Niger, Rwanda, Upper Volta and Zimbabwe.

Zambia was created by the political chess game called "the Scramble for Africa" which was played by the former metropolitan powers. In the words of Jack Simmons, "the Scramble for Africa was what its name implies: a haphazard business of smash-and-grab, conducted with no rules except those of diplomacy, and without attention to the wishes or interests of Africans".[4] The creation of African states, especially land-locked countries like Zambia, was a violation of what is now considered a fundamental right of sovereignty, pursuant to Article 2.4 of the United Nations Charter.[5]

Land-locked countries, the world over are united by a common denominating problem of gaining unimpeded or free access to the sea. Thus, in the period before the 1958 Conference on the Law of the Sea, some twelve non-African countries took concerted action to convene a pre-Conference in Geneva in February 1958. The end result of the pre-Conference included the formulation of an agreed position in respect of the existing law of access to the sea for land-locked countries. In other words, they reiterated the fundamental principle of the Right of Free Access to the Sea as is applicable in respect of freedom on the High Seas.

Proceedings of the Fifth Committee of the United Nations Conference on the Law of the Sea led to the adoption of the

Convention on the High Seas. Article 3 of the Convention reiterates concern for land-locked countries in respect of free access to the sea, namely:

1. In order to enjoy the freedom of the sea on equal terms with coastal States, States having no sea-coast should have free access to the sea. To this end, States situated between the sea and a state having no sea-coast shall by common agreement with the latter and in conformity with existing international convention accord:
(a) To the State having no sea-coast, on a basis of reciprocity, free transit through their territory; and
(b) To ships flying the flag of the State treatment equal to that accorded to their own ships, or the ships of any other States, as regards access to seaports and the use of such ports.
2. States situated between the sea and a State having no sea-coast shall settle, by mutual agreement with latter, and taking into account the rights of the coastal state or of transit and the special conditions of the state having no sea-coast, all matters relating to freedom of transit and equal treatment in ports, in case such States are not already parties to existing international Conventions.[6]

The issue of transit routes is a constant political and economical factor for these land-locked countries. Zambia had no transit route problems during British colonial rule because Britain's oldest ally, Portugal, controlled the lands to the east and west, through which Zambia's traffic passed.

THE UDI'S EFFECTS ON THE ZAMBIAN ECONOMY

It was on 11 November 1965 that one of Zambia's most important transit neighbours unilaterally declared its independence from Britain. Rhodesia's Unilateral Declaration of Independence (UDI) came under the rule of Premier Ian Smith. The move was not totally unexpected by the British government. On the contrary, in the months before the UDI the Prime Minister of the United Kingdom, Harold Wilson, had been holding talks with Mr. Smith in what seemed like a desperate effort to avoid the declaration.[7] Mr. Wilson had even flown into Harare (Salisbury) a month before the action was taken as if to add drama to the episode. The world at large regarded UDI as a duel between Britain and her rebel Colony, Rhodesia. This view was confirmed by the stand adopted by the British Government in claiming that they had sovereignty over the Colony. The Britons claimed that the rebellion would be ended in a matter of weeks, rather than months. As events unfolded, it was Zambia's economy, not Britain's which had to pay the price. Zambia paid a highly disproportionate burden of the cost of UDI because the British Government fumbled in its handling of the affair.

The history surrounding the UDI, especially Zambia's first decade of independence, confirmed the presence of neo-colonialism and illustrated Zambia's economic dependence on outside factors. In the region, the colonial regimes perpetuated neo-colonialism by carving nations into production cells, dependent on metropolitan powers.

Zambia's traditional, most important and reliable route had always been through Southern Rhodesia to the Mozambique ports or those in South Africa. With the UDI and Zambia's adherence to the UN call for sanctions, her traffic on these routes in the succeeding years began to decline. The Rhodesian regime closed its border with Zambia itself on 9 January 1973 as a retaliatory action, for going along with the UN sanctions. Border closure was also due to Zambia's harbouring of the so-called African Rhodesia guerrillas.

Problems associated with weathering the adverse effects of the UDI were perhaps the biggest that faced Zambia in its early independence days. The challenges were political and economic. Politically, one problem was sharing a boundary with a country which was at war with itself and classed as an international outcast. Another political problem was Zambia's adherence to the ideals of the United Nations forcing it to call for economic sanctions against the rebel British colony. The political situation was made worse because Zambia played host to a great many Rhodesian refugees (mostly Africans, and a few run-away whites from South Africa). Zambia also had to prepare for a possible war with the Rhodesian regime. This entailed diverting some of the country's scarce resources from non-military to military use. Military commitments are rarely economically productive.

It is not always possible to obtain a tightly separated definition of what is purely political or economic in terms of many of the problems of land-locked countries. For instance, the legal duties entailed for a land-locked country in respect of exercising the right of access or transit of a land-locked state to respect the rights and interests of transit routes entails the duty to defray all direct costs involved in the provision of transit facilities. The Zambian experience with the use of Tanzanian transit facilities during UDI was such classic example. In this connection, Zambia paid for the construction and temporary operation of the Port of Mtwara on the Tanzanian coast and most of the improvements to the Dar-es-Salaam port.

Another economic challenge resulting from the UDI was finding alternative transit routes comparable in terms of reliability and low per unit cost to the traditional southern routes through Rhodesia and South Africa. Loss of the use of southern routes proved both advantageous and disadvantageous for the Zambian economy.

It is possible to use some of the events associated with UDI to explain not only some of the observed tendencies of radicalism in Zambia but also the lack of it in certain sectors. After the UDI, Zambia made political decisions not in keeping with traditional western capitalist philosophy.

Perhaps a review of some of these decisions will give the reader a better appreciation as to the reasons behind the apparent shift in political thought. Zambia quickly accepted aid from a non-traditional source when China was contracted to build Tazara railway. This was after the IBRD adopted what from the Zambian standpoint could best be described as delaying tactics. Similarly, Soviet military hardware was purchased to compensate for the reluctance of the West to provide Zambia with weapons to defend itself and fight against the Rhodesians. Rhodesia was seen as an extension of South African and Western investment interests. Zambia also showed more open support for the Organization of African Unity and the United Nations in championing the collective decisions of the Frontline States by applying sanctions with immediate effect against Rhodesia. Some of these actions were taken in spite of considerable adverse effects to the Zambian economy. In other words, Zambia had chosen to suffer rather than sacrifice her principles for the economic benefits of fraternizing with the Rhodesian regime.

In the meantime, business enterprises based in the countries where UN sanctions would create economic and industrial problems (like Britain, France and Japan) were openly fraternizing and making profitable business with Rhodesia. Their actions directly and indirectly abetted sanction-busting. Examples included the French Renault and Pegueot car companies and Japan's Datsun motors. Petrol and petroleum products from British, Dutch and French refineries had no difficulty in reaching Rhodesia. Rhodesia's own methane plant only came on stream very late in the history of the UDI. Yet the Zambian economy was starved for petrol. Zambia had to get these products at considerable cost because of the difficulty of rerouting supply lines.

The stage for Zambian economic ills was set during the colonial era. The adverse effects of the UDI on the economy were just more added obstacles to development.

The high cost suffered by Zambia could include some of the never-to-be-recovered investments into the Tanzanian transport and communications networks referred to above.

Zambian policy-makers were further troubled by having to ensure a continuous supply of power from the Kariba Hydropower station on the South Bank, which was under the constant threat of disruption by Rhodesia. It was not without reason to assume that in the war-of-nerves situation between Zambia and Rhodesia that some white Rhodesian extremists would have turned off electricity to Zambia.[8] The economic and the political consequences of such an act would have been disastrous. The first casualty would have been Zambia's lifeline—the mining industry. The country would have not been able to mine and export its minerals, which provided the bulk of the necessary development revenues. The whole economy would have been ruined.

There were other consequences of the UDI on the Zambian economy. Zambia lost access to institutes of higher learning based in Rhodesia, principally the previously jointly-owned university, later named the University of Rhodesia during the UDI era. Its significance is obvious when we recall the very low academic and technical capacity of the Zambian technocrats. Also lost was access to economic facilities which during the Federal days, were located in Rhodesia.

In a nutshell, the history of the UDI was a strange mixture of untold economic hardship on the Zambian economy and brutality with the support of a number of Western economies. It partly explains the reasons for failure of economic sanctions to bring down the regime of Mr. Ian Smith.[9] To all intents and purposes, Zambia became a true expression of one of the Bemba proverbs "of the grass suffering from the trampling of two fighting elephants". This analysis provides a picture of some of Zambia's difficulties with regard to the UDI. The UDI caused unprecedented high costs to the Zambian economy in providing immediate and long term reliable alternative transit facilities. It also caused problems with the supply of essential items which were previously imported relatively easily from across the Rhodesian border.

TRANSIT ROUTES AFTER THE UDI

It was barely a year after Zambia's independence that the Southern Rhodesian government opted for the UDI. At that time, Zambia had not been able to restructure its economy or develop its transport and communications systems to be able to cope with applying UN sanctions which stopped the use of transiting facilities of Southern Rhodesia. The challenge of finding alternative routes was the hardest blow to hit the young nation during its early years. To disengage from the south required enormous resources since Zambia's resources from the export of copper had already been depleted in boosting the Southern Rhodesia economy.

It had become abundantly clear to Zambia that her country's future lay above all else, in finding alternative transit routes. Some railway traffic had to be shifted to the Benguela Railway through Zaire to the port of Lobito in Angola. Traffic was also shifted to road routes connecting Zambia with countries in East Africa like Tanzania and even as far north as the port of Mombasa in Kenya. The Tanzania-Zambia Railway (Tazara) and the Mpika-Mbala road became priority projects.

The development of new routes was an effort on a scale never before imagined because of the big time constraint. Skeptics expected that Zambia would fail because it lacked the resources needed for such a seemingly impossible task. However, the UDI provided, if anything else the needed basis for the process of delinking by Zambia from the southern regimes of Rhodesia with minimum domestic resistance and with maximum international support given to the adoption by the United Nations General Assembly of sanctions against Rhodesia. The Zambian Government therefore adopted a careful policy in protecting the country's national interest.

The policy and the actions taken were not only strategic but also flexible. There were instances when Zambia temporarily soft-pedalled on some of its disengagements. This was necessary because the country had not sufficiently developed alternative transportation means. This is illustrated by Zambia's initiative in the disbanding of Central African Airways (CAA) while disengagement from the unitary Rhodesia Railway system came much later and not even at the height of the conflict between the Ian Smith regime and Zambia as fueled also by the Zimbabwean freedom fighters from bases in Zambia. In its place was created Zambia Airways Corporation with a Zambian, Tom Mtine as Chairman. The Government of Rhodesia could not take retaliatory measures because it realised that such action would be of no significant consequence on the Zambian economy.

The development of new transit routes created enormous problems for the Zambian economy. One was the long period between planning, project implementation and the actual fruition of a project. Another was the mobilization and development of resources in the face of other equally important national demands. The reluctance of the International Bank for Reconstruction and Development (IBRD) to finance the Tanzania-Zambia Railway illustrates the serious financial problems created by this sudden need for new transit routes.

The railway system in Zambia immediately before the UDI had been one of the unitary systems in the region. It was part of the Rhodesia Railways network. Zambia's intention was to develop its own transport industry as part of its post-independence programme (analysed in Chapter Five, along side the emergence of parastatals). The UDI hastened government action in assuming direct control over the international railway system. Assuming control and management of the national railways network was complicated by many factors. For instance, the system relied on the workshop in Bulawayo in Rhodesia for all major repairs and overhaul of equipment. Furthermore, most of the equipment used in Zambia was either obsolete or overused.

However, substantial progress in terms of organization and

rationalization programmes of Zambia Railways had been made. For instance, a new railways workshop was constructed at Kabwe at a cost of K12 million. The railway's infrastructural capacity began to register substantial improvement as the total length of the main and branch lines increased from 827 km in 1972 to 1,104 km by 1976.

The development and improvement of external road connections were of great potential to the Zambian economy. Priority had to be given to this sector because the situation demanded that a proportion of the country's traffic be transhipped through Malawi and then to Mozambique or straight through Mozambique. Goods moved in this direction were transported by road on the Lusaka-Chipata road to Lilongwe (Malawi), via Mchinji, before being moved by rail to the coast. Other goods were moved by road to the railway head at Moatize before being railed to Beira. The same pattern applied for goods moving through Botswana, where transhipment was undertaken at Francistown before being moved to Nata and across the Kazungula Ferry into Zambia, or vice versa.

Several problems arose in the use of these routes. Their use became a test of Zambia's resilience in coping with a totally new high cost situation. These routes offered very limited carrying capacities, which were difficult to increase in a short space of time. For instance, the axle weights on some of railways was very limited. Also, Benguela Railway was rendered unreliable because of frequent distruptions caused by units of the Angola UNITA guerrillas, fighting a war against the Luanda-government in Angola. Problems with the road network revolved around their poor condition and most became impassable during certain periods of the year. Thus the Kapiri Mposhi to Dar es Salaam road was nick-named the Hell Run.

In order to cope with these constraints, the Zambian government made decisions which were absolutely necessary. Zambian policy-makers decided to airlift bulk cargo, such as copper and fuel, to and from Indian Ocean coast ports. The country's resources were overstretched because the country had to continue with most of its other national development projects.

To a large extent, the airlifting of copper and fuel was made possible because of external assistance. Also important was the country's own financial strength in meeting a major part of the cost involved. Another help was the goodwill shown by the

Zambians themselves toward some of the measures taken by the government. In particular, they accepted such hardships as petrol rationing and going without certain essential items.

The airlifting episode provided for income generation, with a multiplier effect for the majority of the people, through greater utilization of domestic resources. It also offered an opportunity to diversify the production of industrial and consumer goods. Part of this diversification was constructing the Tanzama Oil Pipeline from the Tanzanian port of Dar es Salaam to the Indeleni Refinery at Ndola. Other diversification efforts were the development of the country's hydro-power station at the Kafue Gorge and the opening of the coal mines at Maamba.

The problems brought about by having to reroute traffic after the UDI went beyond cost considerations. Naturally, the cost of construction was high and Zambia's own topography increased costs. Another issue was the uncertainty about the mood and reaction of the neighbouring states who were to help provide the alternative means of transport and transit facilities. The general objective was therefore to evolve a comprehensive transport strategy whose general objective was to secure an unimpeded flow of goods at the minimum cost possible. These were serious questions to which Zambian policy-makers had no ready answers.

Consequently, one of Zambia's challenges was to persuade the transit states to accept the development of transit facilities as joint ventures with mutual benefits. This was the reasoning behind the tarmacing of the road between Kapiri Mposhi in Zambia and Dar es Salaam in Tanzania. And the construction of the railway connecting Zambia and Tanzania (done with the help of the People's Republic of China, as a tripartite project). The same reasoning was attempted in getting agreement between Zambia and Botswana and Malawi on joint road projects. However, progress was slow with regard to arriving at an agreement with Botswana.

After the UDI, Zambia was in a situation where she had no alternative but to develop alternative routes. Perhaps Zambia needed the impetus of the UDI to discover her own economic potential and to initiate developments that concentrated in areas other than mining. The effects of the UDI also brought home some of the realities of associating closely with other independent African countries outside the usual pattern of African political goodwill. The resolute application of Zambia's determination to survive economically against the odds presented by the UDI provided others with some telling lessons. Some friendly neighbours thought they could take unbridled advantage of the situation. As it turned out, Zambia's strong financial position formed a major influencing factor on the transit states.

By weathering some of the storms created by the UDI, Zambia's original economic weakness, dating back to colonial and federal days, became a source of strength. It can only be hoped that in future, Zambia would never again allow any one of its transit neighbours to use its weakness as a source of their strength to undermine Zambia's goals of economic development.[10] Consequently, the country learned that it could face a tremendous challenge and be strengthened by its successful handling of that challenge.

REFERENCES

1. W.E.F. Ward, *Emergent Africa*, George Allen and Unwin Ltd., Norwich, 1967, p. 77.
2. Jack Simmons, *Livingstone and Africa*, The English Universities Press Ltd., London, 1966, pp. 19-152.
3. Zdenek Carvenka, (ed.), *Landlocked Countries of Africa*, The Scandinavian Institute of African Studies, Uppsala, 1973, pp. 188-193.
4. Jack Simmons, *op.cit.*, p. 161.
5. Article 2-4 of the *United Nations Charter* states that "All Members shall refrain in their international relations from the threat or use of force against the territorial integrity or political independence of any state, or in any other manner inconsistent with the Purposes of the United Nations".
6. United Nations Conference on the Law of the Sea. *Fifth Committee*, Geneva, 24 February—27 April 1958, Document A/CONF. 13/43, Volume VII, Footnote no. 1, p. 33.
7. David Smith and Colin Simpson with Ian Davies, *Mugabe*, Pioneer Head (Pvt.) Ltd., Salisbury, 1981, p. 62.
8. Paul Moocraft, *A Short Thousand Years: The End of Rhodesia Rebellion*, Galaxie Press, Salisbury, 1980, p. 21.
9. Ibid., pp. 180-194.
10. Henry Bretton, *Power and Politics in Africa*, Longman, London, 1977, pp. 55-57.

Chapter Three
Development Planning in Zambia

For people who have been slaves or have been oppressed, exploited and disregarded by colonialism or capitalism, "development" means "liberation". Any action that gives them more control of their own affairs is an action for development, even if it does not offer them better health or more bread.[1]

A GENERAL OVERVIEW OF DEVELOPMENT PLANS

More often than not, planning in developing countries is discussed on two planes. On one hand, substantially as a technical problem by technicians who claim to be experts of conventional theories of development. These planners tend to emphasise the sectoral approach. On the other hand, is the discussion of planning as mechanical problems. Concern in this case lays emphasis on resolving the administrative machinery but which other experts apart from the administrators are also expected to provide solutions. However, these two approaches need not be mutually exclusive. The analysis which follows in this chapter will shed a little more light on many of these problems.

According to the United Nations Economic Commission for Africa, "development planning (in Africa) has become an accepted method of ordering the economic and social lives of the people. The planning idea is not entirely new ...but it certainly gained in popularity during the period of struggle for political emancipation. Popular political parties had been compelled to draw attention to the numerous economic and social disabilities imposed by foreign rule, and even when they published no reform programmes, they left the people in a state of joyful, but impatient expectancy.... The popularity of planning was also enhanced by the programme of the United Nations ...declaring the decade of the sixties as the Development Decade....: Further, it urged the adoption of planning to ensure orderly progress....."[2]

This thesis discusses development planning only as it relates to economic development, because the subject of development planning in general is already a much discussed topic. Specific information and its relevance or applicability is reviewed against the Zambian experience, along side the overall experience of the other Third World countries. The central theme analyses post-independence issues of economic growth and development planning.

Basic to understanding development options of the Third World countries is knowledge of what unifies a society and the establishment of its government. Similarities in social and economic problems confronting developing African and the other Third World countries account for their similar approaches to development planning. The types and main features of their development plans and programmes show fairly similar targets. For the majority of developing countries analysed herein, there has been no unified approach to development planning. Rather, there has been a need-oriented and self-reliant development, geared towards more endogenous and innovative processes. These take into account environmental limits and potentials and anticipate greater participation of the people as a whole.

The struggle and eventual attainment of political independence for many formerly colonized states gave rise to many pious hopes: to promises of a better and more abundant life for all and of a just and egalitarian society; to visions of a bright future and greater job and education opportunities for many; and to an improvement in the level and quality of services provided for the welfare of the nation at large. For many developing countries the study of economic development is a relatively recent phenomenon. The analysis of economic development is by its very nature multi-dimensional and an on-going process. It draws heavily on the country's history. It also differs, as will become clear later on, from the analysis of economic growth, even though the two are not mutually exclusive.

The development plans recognise the paramount importance of raising living standards of the masses. However, developing countries' successes in adjusting to new socio-economic and political environments have been few because of the countries' lack of resources with which to cope with the world economic crisis.

Development Planning in Zambia

The accompanying table of development plans and programmes in Africa into the 1970s is illustrative of the range and character of those throughout the Third World. No less than 40 African countries had attained political independence by 1968. Approximately 85 development plans or programmes were conceived and implemented or reformulated by that time. (Except for Ghana, which in 1961 abandoned its 1959/64 programme) They range from a simple statement of intended government spending to the comprehensive coordination of policy objectives and the identification of policy instruments based on systematic models of their respective economies.

The fact that so many development plans had been drawn up in so short a time can be taken as another indication of the African faith in the usefulness of development plans to express their goals or objectives. The Zambian case study may help to throw some light on some of the issues of economic development. This analysis deals with questions of: precise definitions; specifications and the direct relationship between the instruments of development (such as investment) and the real objective (such as the improvement of living conditions through the identification) of needy groups; the use of both transitional and perspective methods as bases for medium-term and short-term plans; and the establishment of ongoing research as an integral part of the formulation and implementation of plans.

Table 3.1 shows that serious planning in 60 per cent of African countries at post-independence tended to begin with the formulation of a transitional plan or *plan interimaire*. Transitional plans usually varied in duration from one to two years. However, there were a few cases where transitional plans had a life span longer than two years. For instance, in Benin, Senegal, Niger, Botswana, Kenya and Malawi (3 years), in Ethiopia and Upper Volta (4 years) and in Egypt (5 years).

Some reasons can be deduced to support a transitional plan for a formerly colonized country. They include, *inter alia*, the need for bridging the period of constitutional transition from internal self-rule to full independence and covering the period of change from the rudimentary planning represented by a colonial public expenditure programme to the full resource planning being introduced by the new government.

The table provides some useful data in respect of the country

TABLE 3.1: Summary of African Post-Independence Development Plans and Programmes

Independence Year	Country	Transitional plan (period)	Development Plan/Programme I (period)	II (period)
Ages	Ethiopia	1958-62	1963-67	1968-73
1922	Egypt	1960-65	1965-70	1973-82
1947	Liberia	—	1951-56	1951-60
1951	Jamhyaria (Libya)	—	1963-68	1972-73
1956	Morocco	—	1960-64	1965-67
	(a) Sudan	—	1961/62-70/71	1970-75
	(b) Tunisia	1962-64	1965-68	—
	(c) Ghana	1957-58	(i) 1959-64 (abandoned 1961)	—
1957			(ii) 1963/64-69/70	
1958	Guinea	1960-62	1964-71	1970-71
1960	(a) Benin	1962-65	1966-70	—
	(b) Cameroon	1961	1961-65	1966-71
	(c) C.A.R.	1960-62	1964-65	1967-70
	(d) Chad	1964-65	1966-70	1971-80
	(e) Congo	1961-63	1964-68	—
	(f) Gabon	1963-65	1968-70	1971-75
	(g) Ivory Coast	1962-63	1965-70	1971-75
	(h) Madagascar	1960-62	1964-68	—
	(i) Mali	—	1961-62	1961-65
	(j) Mauritania	1960-62	1963-66	1970-73
	(k) Niger	1961-63	1965-74	n.a.
	(l) Nigeria	—	1962-68	1970-74
	(m) Senegal	1961-64	1965-69	1969-73
	(n) Somalia	—	1963-67	1971-73
	(o) Togo	—	1966-70	1971-75

	(p) Upper Volta		1963-67	1966-70	1972-76
	(q) Zaire		—	1965-69	—
1961	(a) Sierra Leone	—	1962/63-71/72	—
	(b) Tanzania	1961/62-63/64	1964-69	1969-74
1962	(a) Algeria	1963-64	1967-69	1970-73
	(b) Burundi	—	—	1968-72
	(c) Rwanda	—	1963-65	1964-70
	(d) Uganda	—	1966-71	1971/72-75/76
1963	Kenya	1957-60	1965-70	1970-74
1964	(a) Malawi	1962-65	1965-71	1971-80
	(b) Zambia	1965	1966-70	1972-76
1965	Gambia	E/F*	1964-67	1967/68-70/71
1966	(a) Botswana	1966-69	1970-75	n.a.
	(b) Lesotho	—	1979-75	n.a.
1968	(a) Mauritius	—	1971-75**	n.a.
	(b) Swaziland	1969	1968/69-72/73	1973-77
Grand Total	40	24 (60%)	39 (97.5%)	29 (72.5%)
	Period profile	1 yr × 6	1 yr × 2	2 yrs × 2
		2 yrs × 9	2 yrs × 4	3 yrs × 5
		3 yrs × 5	3 yrs × 3	4 yrs × 10
		4 yrs × 2	4 yrs × 16	5 yrs × 8
		5 yrs × 2	5 yrs × 8	6 yrs × 1
			6 yrs × 2	9 yrs × 2
			7 yrs × 1	10 yrs × 1
			9 yrs × 2	
	Average:	3 yrs.	4.6 yrs.	Average: 4.57 yrs.

Source: Compiled from national data in ECA Library and Surveys of Economic Conditions in Africa by ECA.
** Travail pour Tous (work for all)—more or less to prop up back-log of unemployment and match employment creation with number of new entrants to the labour force.

adherence to development planning models but not much with regard to some glaring shortcomings between the plans in the pre- and immediate post-independence periods in developing countries. The analysis which follows will address issues and difficulties encountered in the choice of growth targets and mechanism for monitoring progress. Numerous studies by the United Nations, especially the Economic Commission for Africa, reveal that the designers had difficulty in choosing between the Western and non-Western schools of thought. The former owes its evolution to the writings of economists like Walter Rostow. The latter owes its evolution to the emerging challenge by economists in Third World countries to the traditional theories of economic development.

The Western (Rostovian) view of development lays stress on the fact that poor countries are poor simply because they are poor. In other words, the main cause of under-development in the developing economies emanates from a lack of capital, which is an important cooperant factor of production. It further assumes that poor developing economies are locked up in vicious circles of poverty from which they are unable to extricate themselves. To get out of this situation, developing countries are advised to adopt the so called *Big Push* of heroic dimensions. These poor countries must rise from poverty and thrust themselves into the capitalist orbit of the developed countries. Some of the means by which this *Big Push* could be brought about include the infusion of massive amounts of foreign aid, massive foreign private investment and transfer of technology from the industrial North to the impoverished South.* Advocates of this line of thinking suffer from illusions of the old economic order. For instance, they are not willing to face up to the realities of North-South relations. They don't appreciate that aid is a cost which the recipient must repay, either immediately or in the long run. Furthermore, they don't appreciate that aid is given for priorities determined by the donor, rarely by the recipient.

On the opposite side is found the school of economic thought championed by Raul Prebisch and the other Third World economists. This group has discovered the centre-periphery theory of world economic relationships wherein the developed countries

* See Chart of Global Interdependence Channels

Development Planning in Zambia

constitute the centre and the developing countries are relegated to the periphery. Walter Rodney has shown that developing countries were perpetual losers in such international economic relations. Neither the Big Push nor the centre-periphery theory has the means of removing the inequality between the developed and developing countries. The Prebisch-Rodney group sees the less developing countries (LDCs) as satellites of the developed countries (DCs), who constitute the *metropoles*. The satellite-metropole relationship is one of dependence of the former on the latter. It ensures the exploitation of the LDCs by the DCs through trade, aid, and the activities of transnational or multinational corporations (TNCs or MNCs) and constitutes the backbone of the current international economic order (CIEO). It owes its origin to the colonial past as it guarantees the continued transfer of resources and wealth from the LDCs to the DCs.

Some proponents of the satellite-metropole theory feel that the way of escape for LDCs from this syndrome is through changes in the structure of the old economic order that would keep LDCs out of the orbit of advanced industrialized countries. Some of the tenets of the New International Economic Order (NIEO) allude to this but it is too early yet for the LDCs like Zambia, to assume that the NIEO is any panacea for its economic challenges and dilemmas.

Some attempt was made in this chapter to relate the virtues of the market mechanism as a means of allocating resources and promoting growth and development. The market mechanism in pre-independence Zambia did not result in a fair distribution of resources and benefits. The new government knew both its challenges and its role in society.

Zambia emerged at independence with a weak economy and with policy-makers who had lost faith in the efficiency of the market system. This is because the market mechanism had failed Zambia. It had crystalised the population into three groupings: Europeans on top and in command of the country's major economic sectors, such as mining and commerical farming; in between, Asians and other non-African minority communities who controlled most medium-level industrial and commercial activities such as wholesaling and retailing and, at the bottom the indigenous African population who were engaged in selling their cheap labour to the two economically powerful groups or were living simply as peasant farmers or small-time traders.

Political scientists like Jean Jacques Rousseau believed that governments are a product of civil societies created for the purpose of assuring the property, life, and liberty of each society member. Governments are created for the protection of all. John Mill urges governments to take action when the situation so demands, for example when the private sector has proved itself incapable of taking self-correcting measures in the interest of all. Development planning fits into Mill's concept that government take action. However, the colonizing countries who rose to power when mercantilism and Adam Smith's "invisible hand" held sway, were not over enthusiastic about planning in their respective colonies. The history of such schemes as the Colonial Development and Welfare Acts for most former British colonies, and the Fonds d'Investissement pour Development Economique et Social des Territories d'Outre-Mer (FIDES) in the French colonies, dealt only with the rudiments of economic planning.

The elements of development plans deal with the efficient allocation of available resources, economic growth through increased productivity, economic development as shown through full employment, fair income distribution, stability and balanced foreign trade. The International Labour Organization defines economic development as meeting a country's *basic needs,* such as attaining full employment, reducing income inequities and eliminating poverty.[3] Of paramount importance is raising the overall standard of living of the people.

At least four categories of development plans have emerged in Africa: (*a*) project-oriented plans; (*b*) plans based on the impressionistic use of national accounts data; (*c*) systematic plans based on national accounts data and sector analysis; and (*d*) plans which were either based on, or made explicit use of formal models.[4]

The prime objective of a development plan has been to provide a framework for change, taking account of the political setting. Issues of importance stressed in the plans have included devising ways of managing a country's major productive sectors, principally, agriculture, mining, industry, and human resources. This implies an internalization of the country's development strategies at the national level. Development plans have provided a framework for reversal of certain trends considered detrimental to a country's long-term plans. For example, Zambia stopped the outflow of native populations as indentured labour to the South African gold mines and Southern Rhodesia through the Witwatersrand Native Labour Association (WENELA). That loss of human resources had its most adverse economic effects in the Western Province, with the depletion of the male population needed for subsistence farm production.[5]

A development plan is therefore a blueprint and deliberate and purposeful act to influence economic activity. Consequently it must contain both desired objectives and methods by which those objectives can be secured.

HISTORY AND TRENDS OF DEVELOPMENT PLANNING

The history of development plans in Zambia is relatively short. What planning that took place before was not done to obtain an overall balance. The Northern Rhodesia Ten-Year Plan, from 1947 to 1956, was an uncoordinated affair. It was characterized by a strong bias for sector development and in particular, those sectors under European and expatriate control.

The year (1963) just before the attainment of political independence was singularly important. It helped to determine: What was to be the main objectives of the country's development plan? What would be the consequences of the disintegration of the Federation of Rhodesia and Nyasaland?[6] Planners turned to the United Nations and its agencies for advice on what to emphasize.

Each agency offered help in its field. The United Nations Educational, Scientific and Cultural Organization (UNESCO) stressed aspects of education; the International Bank for Reconstruction and Development (IBRD) placed emphasis on communication infrastructure. The joint team of the United Nations Economic Commission for Africa (ECA) and the Food and Agriculture Organization of the United Nations (FAO) covered a much wider field. Its terms of reference included planning for a broad framework for social and economic development and to devising suitable development policies.

No less than four development plans had been put into operation in the period under consideration. The first was the Emergency Development Plan (EDP), followed by the Transitional Development Plan (TDP), the First National Development Plan (FNDP) and the Second National Development Plan (SNDP).[7]

The first post-independence plan was the Emergency Development Plan (EDP). It was a prelude to the 18 months Transitional Development Plan (TDP) which ran from January 1965 to June 1966.

The Transitional Development Plan (TDP) was basically a means for bridging the period of transition from internal self-rule to full independence. Secondly, it covered a period of change from the rudimentary planning represented by colonial public expenditure programmes to the initiation of full resource programming, which the Government wanted to introduce. The first category of projects considered were those being carried over for completion. The second group was in two sub-categories. The first sub-group consisted of projects on which work would be started immediately upon the completion of financial agreements. The second sub-group was comprised of new projects whose starting period was contingent upon progress in the two categories already described. In the third category were non-priority activities, such as promoting cooperative societies. The total estimated cost of TDP was approximately K90 million.

One unexplained fact with the scale of resource allocation under TDP was the rather meagre allocation to agriculture relative to other sectors like transport and mining given the need to increase food self-sufficiency and the development of other natural resources. Perhaps, Zambia was after all unlucky to have been born with a copper spoon. For how else, could one explain the low priority given to agriculture and the rural sector, except that revenues from copper continued to comfortably cushion the economy. Perhaps as economic analysis goes, it is usual to justify such action on the basis of paying more weight on the propulsive force of the mining sector than that of other sectors.

A new chapter began to emerge with the preparations for Zambia's First National Development Plan (FNDP). The Office of National Development considered the FNDP, covering the period 1966 to 1970, as a definite departure from previous patterns, not so much in terms of the mechanisms to be used, but rather in terms of the objectives sought.

The objectives were noble and lofty. Some attempt was to be made to involve operatives in certain sectors of the community through the creation of local development committees, but mainly those in the regional centres. A national convention was called to discuss the FNDP in Kitwe from 11 to 15 January 1967. Just convening of the convention opened a new chapter in Zambian history. Architects of future plans should have the courage to draw lessons from some of the Convention's conclusions.

The following list of goals and areas selected for in-depth discussion at the convention provides the reader with an idea of the sort of emphasis placed on the various sectors:

(*i*) diversification of the economy to eliminate over dependence on the copper industry and to encourage the establishment of the wider range of other industries, in order to reduce reliance on imported consumer goods;

(*ii*) increase employment by at least 100,000 jobs during the course of the plan;

(*iii*) increase the value of the nation's own output from K122 to K200 per annum per head by the end of the Plan period;

(*iv*) maintain reasonable price stability;

(*v*) change the situation so that the urban population could no longer command a greater part of job and wealth opportunities at the expense of the rural population;

(*vi*) rapidly raise the level of general education and provide training to equip more people for administrative and technical jobs in the professions and as managers of all sorts of enterprises from cooperatives to workshops or businesses;

(*vii*) provide the necessary social infrastructure, better houses and other social facilities, to raise the standard of living of the people; and

(*viii*) develop new sources of energy, transport and communications and other physical infrastructure which constitute the basic framework for real economic growth and development.

Data in Table 3.2 pinpoint areas of concentration by the public sector during the FNDP. A large proportion of public expenditure was in the transport and communications sector. It accounted for approximately 22.9 per cent of the total public sector investment of K564.6 million and 15.1 per cent of the total national investment of K858.6 million. Housing and construction, education and agriculture absorbed about 15.3 per cent, 14.19 per cent and 12.2 per cent respectively. Bottom of the scale were forest, game, fisheries and research to which were allocated 3.2 per cent notwithstanding

TABLE 3.2: Public Investment by Economic Sector (FNDP) 1966—1970
(Million Kwacha)

	Public Sector	% Sector	As% of Total
Agriculture	69.0	12.2	8.0
Mining	41.0	7.3	4.8
Manufacturing	50.0	8.9	5.8
Forest, game, fisheries and research	18.0	3.2	2.1
Power (electricity)	53.4	9.5	6.2
Transport and communications	129.3	22.9	15.1
Housing and construction	86.3	15.3	10.1
Education	79.4	14.1	8.3
Health	18.4	3.3	2.1
All others	59.8	10.6	6.9
Sub Total	564.6		65.8
Nation Total	858.6		100.0

Source: Derived from data given in *Zambia's Plan at Work*.

the importance of research to a country's development process. The public sector shouldered over 65 per cent of the nation's total investment. Most investment was channelled through the parastatals, such as Indeco.

The manufacturing sector at that time concentrated on import substitution because of the need to cope with the effects of the UDI. Import substitution in economic theory implies domestic production of goods that would otherwise be imported. It is therefore important to review industrial development as a strategy in the country's process of economic development and also the way it was viewed in the various development plans. The pattern of import substitution or industrialization that emerged in the Zambian context was nationally conceived due to the pressure brought to bear on the economy as a direct consequence of the UDI. Perhaps this is why the strategy failed to provide for maximization of backward and forward linkages in the economy. Production by the respective industries also did not provide for horizontal and vertical integration.

The result was that old structures persisted for too long. For instance, many industries did not use enough local raw materials in the manufacturing of their products. Industrialists rejected local inputs becaue they considered them to be of inferior quality. When because of the UDI and Rhodesia's closure of the Zambia/Rhodesia border coke from Wankie colliery could not be obtained for the Zambian mines, the Zambian copper mining technocrats continued to resist to adopt the use of Zambian coal from Maamba mine. This could have been the ideal opportunity for adaptation had research not been given such a very low priority. Consequently, this type of industrial strategy simply increased local demand for imported intermediate and capital goods. The same pattern also appeared in the brewery industry where the import of barley, etc., was not reduced. Whatever hurried import substitution investment that went on, merely reinforced the old income redistribution channels. It is the paradox in such a situation that the income generated by such investments remained mainly in the urban areas, where the industries are located and went to the high income earners in society.

The Second National Development Plan (SNDP) was supposed to continue the work of the FNDP. Its goals were to increase and expand the diversification process initiated under the FNDP in order to reinforce:

(*i*) attaining self-sufficiency in food supplies together with an improvement in income;
(*ii*) expanding and diversifying industry and mining through import substitution, which emphasised a greater utilization of local materials and ensuring an improvement of living conditions in the rural population by creating employment;
(*iii*) initiating comprehensive measures for regional development;
(*iv*) linking educational programmes to the country's manpower requirements;
(*v*) providing social infrastructure on a country-wide basis; and
(*vi*) providing the necessary development of physical infrastructures such as transport and communications; power, etc.

Repeated reference has been made to the historical pattern of

African economies as exchanging primary export for manufactured goods with the metropolitan industrial powers. This left countries like Zambia with a great need for industrialization. Political independence and the UDI, for Zambia, was construed *interalia* as an occasion for breaking with the sad past. Thus, a development plan usually emphasises industrial growth associated with import substitution, increased processing of primary products and export diversification. However, we shall examine later in greater detail the progress and problems of industrialization in the post-independence era in Zambia.

Data in Table 3.2 show that it was the public sector which again shouldered nearly 65 per cent of total sectoral investment under the SNDP, as was the case under the FNDP. Sectorally, public investment in the transport and communications infrastructure continued to receive top priority. It was increased from

TABLE 3.3: Investment by Economic Sector* (SNDP)
1972-1976
(Million Kwacha)

Activity Area	Public Sector	Per cent	Private Sector	Per cent	Total
Rural sector	122.5	9.6	30.9	4.4	152.5
Mining	41.0	3.2	361.0**	52.7	402.0
Manufacturing	107.0	8.4	146.0	21.8	253.0
Construction	—	0.0	63.0	9.2	63.0
Power	198.8	15.6	—	0.0	198.8
Trade	45.0	3.4	10.0	1.5	55.0
Transport and communications	369.3	29.1	5.4	0.8	374.7
Tourism	15.5	1.2	10.0	1.5	25.0
Housing, etc.	146.0	11.5	54.0	7.9	200.0
Education	114.5	9.0	3.0	0.4	40.0
Health	37.0	2.0	3.0	0.4	40.0
Others	74.9	5.9	—	0.0	74.9
Total	1,271.0	100.0	685.4	100.0	1,956.4

* *Source*: Compiled on the basis of data given in the Table 1.1 to Table 3.2 of Total Investment Outlay in the Second National Development Plan (1972-1976). Ministry of Development Planning and National Guidance, Lusaka, Zambia.

** Mainly investments of the subsidiaries of MINDECO which are 51 per cent owned by the government.

22.9 to 29.19. Investment into most other economic sectors except for generating electric power, declined. Notwithstanding information pertaining to mining in Table 3.2, the private sector concentrated its investment about 21.3 per cent of the total, in the manufacturing sector.

Zambia's experience with using development plans to bring about desired economic development confirms the fact that newly independent countries had many obstacles to overcome.

A review of Zambia's development plans, especially the way they were executed, reveals some glaring shortcomings. For instance, there is a persistent lack of co-ordination with regard to the various sectors. Change in the industrial sector does not seem to have gone on apace with that in the agricultural sector. The extent to which some of the recommendations made at the Kitwe National Convention were actually realized is difficult to establish. Successive development plans continued their over reliance on the public sector underwriting the bulk of the country's investment. Efforts at mobilizing the indigenous population or entrepreneurs to feel themselves a part of the whole exercise were perhaps the most difficult work attempted in the plans. This explains why there is a lack of balance in development between the rural and urban sectors of the economy, even after a full decade of independence.

RESULTS

Regardless of form, the decision to plan for economic and social development generally implies an understanding of most characteristics, constraints and potentialities of an economy. It also implies a deliberate and conscious decision to manipulate the economy on the basis of relevant knowledge. Answers to the following questions provide insight into the requirements for effective development planning. First, were Zambia's development plans historically specific, i.e., did they specify targets and policies which could be said to be plausible under prevailing conditions? With the above could also be raised the issue of administrative capacity, since implementation of any plan is a government responsibility. Secondly, were these plans flexible and adaptable to regular revision? Did they give an expression to social, political and economic aspirations?

Data in Tables 3.2 and 3.3, together with that in Table 3.4, is fairly illustrative of the progress made in the Zambian economy during the period under review. The aspect of growth in production in Table 3.3 deals with agriculture, industry and services against the background of the GDP. Additional information is provided in Tables 3.4B, 3.4C, 3.4D and 3.4E on the structure of merchandise trade, the balance of payments and the debt services ratios, the external public debt and international reserves and the flow of external capital, respectively.

Certain conclusions can be reached on the basis of data in Table 3.4, against the backdrop of the GDP between the two periods. The average annual growth rates during the SNDP period for both agriculture and industry is in marked contrast with the service sector, which experienced a decline from 8.1 per cent between 1960 and 1970 to 4.1 per cent for the 1970 to 1976 period. The rather disproportional increase in the balance of payments and debt service ratios was a combination of a high imported fuel costs and payment for infrastructural commitments necessitated by the rerouting of transit routes. The need to provide social infrastructure, such as education and health on which to build future economic development, also contributed to the worsening balance of payments. This is so because investment in such social sectors has a very long period before producing dividends. However, the data in Table 3.3 are not the only criteria by which to judge the Zambian experience. Development that benefits people must also be judged by how it makes the people participate in considering planning and implementing their development plans.

The significance of the transport and communications infrastructure to economic development has been reviewed in Chapter Two. However, a deeper appreciation of data in Table 3.4 requires additional sectoral details. In this connection, the contributary aspect of a rise in the external public debt, especially that necessitated by the need to cope with recurrent expenditures, is worth noting. This tended to outstrip the budgeted figure by a considerable margin. It rose nearly three-folds between 1965/1966 and 1973 from K139 million to about K395 million.

Under the SNDP, it was decided to construct the joint Tanzania-Zambia railway (Tazara).[8] This railway was intended to spearhead and assist with other economic developments. With these developments emerged such ideas as the establishment of Intensive

Development Zones and giving the agriculturally rich Northern Province an opportunity to exploit its potential and to reverse the rural depopulation trend prevalent during the colonial era.* The SNDP also provided for considerable activity connected with the expansion and development of the country's trunk roads, connecting the main highway systems with neighbouring countries. Steps were also taken in improving district, rural, branch and feeder roads in order to provide for an integrated development of the economy.

It was realised by the Government that a well-run roads and railways system could bring considerable savings in time and efficiency in distribution. The paving of the Princess Nakatindi Road linking Livingstone (in Southern Province) and Mongu (in Western Province) resulted in a substantial time reduction (seven hours) for travellers between Livingstone and Sesheke.

TABLE 3.4: Sectoral Economic Trends

A. *Growth of Production: Average Annual Growth Rates (per cent)*

	1960-70	1970-76
GDP	4.0	3.1
Agriculture	2.0	3.2
Industry	0.1	3.4
Services	8.1	4.1

B. *Structure of Merchandise Trade (Exports and Imports)*

	Percentage Share of Exports			Percentage Share of Imports	
	1960	1975		1960	1975
Commodities		99	Food	—	8
Manufactures	0	1	Fuel	1	14
			Other	—	78
Total		100		—	100

* Hans Hedlund and Mats Lundah, *Migration and Change in Rural Zambia*, The Scandinavian Institute of African Studies, Uppsala, 1983.

C. *Balance of Payments and Debt Service Ratios*

	(US$ million)		(Per cent)	
	1970	1976	1970	1976
Current account balance before interest payment on external public debt	131	−571		
Interest payment on external public debt	23	52		
Debt service as percentage of				
(i) GNP			3.2	2.4
(ii) Export of Goods and Services			5.4	8.9

D. *External Public Debt and International Services*

	(US$ million)		(Per cent)	
	1970	1976	1970	1976
(a) External public debt				
(i) Outstanding and disbursed	548	1,184		
(ii) As percentage of GNP			32.0	53.7
(b) Gross international reserves				
(i) Amount	514	100		
(ii) In months of import coverage				10.8

E. *Flow of External Capital*

	(i) Public Guaranteed Medium- and Long-Term Loans (US$ million)		(ii) Net Direct Private Investment (US$ million)	
	1970	1976	1970	1976
Gross inflow	335	270		
Repayment of principal	31	45		
Net inflow	304	225	−297	

Source: *World Bank Annual Report* 1978.

However, transport and communications infrastructure, to be of major economic value, has to be supported by greater usage in both directions. It also requires a tariff policy that allows for reasonable transport rates in both directions. Zambia's main manufacturing industries are mostly found in the few urban centres. Therefore, optimum national use of the communications network demanded a rapid development and activation of the country's other resources, especially those in the rural areas.

The construction of an all-encompassing telecommunications network as well as an internal air transport system received reasonable attention during the SNDP period. The inland water system in addition to the already existing Zambezi River Transport (ZRT) also received due consideration. Attempts were made to economically exploit the potential of beasts of burden, such as donkeys and mules, in transporting goods in certain rural areas. Unfortunately, this line of thought did not go beyond the drawing board.

Other major developments on which attention was focused in the plan included the provision of electricity to most parts of the country. Modern economic activities require electricity to complement other developments. The manufacturing industries and most of the communications networks listed above would have encountered considerable setbacks had an adequate and reliable supply of power not been sufficient and readily available.

Zambia's other important economic sector, apart from mining is agriculture. It is vitally important in the context of the country's overall strategy for economic development. It is the most important single sector in the economies of most African countries except those like Zambia where minerals contribute a far much higher proportion of the gross domestic product. However, the importance of minerals does not diminish the potential of the agricultural sector as can be seen even for the oil rich countries like Nigeria. Agriculture is a major employer of rural populations in the less industrialized countries. It also provides a market for industrial products and for the non-agricultural sector, with industrial raw materials. It should feed not only the rural populations but also dwellers in the industrial urban areas. Empirical evidence shows that only at a much later stage of economic growth and development does industry play a role comparable to that originally played by agriculture.

Through the various development plans, the Zambian authorities attempted to develop agriculture. Government expenditure on agriculture infrastructure included provision for, among other things, water wells, canals, feeder roads, collection depots, partial electrification, research in better livestock and the establishment of agriculture research and training institutions. The most difficult task was to transform traditional agriculture into modern agriculture. Traditional agriculture is barely more than subsistence production and includes the use of primitive technology.

However, the lack of clarity in agriculture policy may have accounted for some of the competing strains in production priority for certain commercial primary exports like tobacco, maize and groundnuts. Many of the rural areas failed to produce enough food to feed themselves during the latter part of the SNDP. A quick reference to data in Table 3.4 shows a much higher average annual growth of agriculture during the period 1970 to 1976. Some agricultural problems should be considered alongside the country's tax structure. No tax considerations were provided to farmers in the plans. Perhaps the root cause for agriculture failures was the way the technocrats at the Ministry of Agriculture went about drawing up the strategy for agriculture development.* One major point which seems to have been overlooked was establishing the importance of demand and supply constraints. Creating of favourable market opportunities could have induced adequate supply response from the farmers. Similarly, unfavourable market opportunities could have been overcome by an exceptional supply response through proper incentives and good producer prices to farmers.

CONCLUSIONS

This chapter has thus far described an experience in economic development planning. There is ample evidence that in Zambia, as in many other developing African countries, there was a definite lack of a follow-up system and operational control of a plan once the plan was launched.

* It is evident that the production of cash crops was given greater prominence unlike food self-sufficienty. Cash crops were favoured because of the inherent foreign exchange they could earn and were therefore produced on fertile lands. On the contrary, the production of food crops was mostly on marginal or non-arable lands. Thus, the subsequent food crisis is partly a result of their short-sighted policy.

Development Planning in Zambia

The method adopted in Zambia in drawing up the First National Development Plan represented an advance in comparison to other African countries who were in similar positions. On the basis of available data and experience, Zambia's method was comparatively better than say Somalia and Ethiopia. Countries who are supplied with good statistical information could greatly benefit from emulating the Zambian example.

In a nutshell, the Zambian method involved, initially, the collection of resource data, thereafter, inviting ministries and provincial authorities to submit imaginative maximum programmes. These programmes were set out on project data sheets giving the details of necessary inputs. The plan had a 42-column input-output table as compared to Mali's 8 (in the Mali Plan), Tunisia-17 columns, Algeria-27 columns (in the Perspective Plan), Morocco—30 columns and Egypt—33 colums (in the Perspective Plan) and 83 columns (in the Five-Year Plan). In addition, Zambia's projections of national accounts for the terminal year (1970) were prepared *inter alia* on the basis of export volumes and prices, rates of investment, foreign exchange, growth in monetary output of agriculture, and employment etc. It was also assumed that the plan would be kept under review throughout its lifetime.

There were innovations in Zambia's economic reform process. A national conference was convened in Kitwe to discuss the implementation of the Second National Development Plan. There was also the idea of creating "Intensive Development Zones". The zones were to concentrate public services and investments in order to promote the integrated development of rural areas. The establishment of these zones was to start in areas with the greatest inherent potential within the provinces. Their aim was basically to: decrease social and economic disparities between the rural and urban areas; help create self-sustaining improvement in incomes and productivity in the rural areas; moderate the rural-urban rate of migration (see aspects of the population in Chapter One); contribute to national self-sufficiency in agricultural commodities; build a foundation for the future development of rural areas.

The analysis has covered the framework of objectives and the difficulties of translating blueprints into practice, the extent to which objectives had actually been implemented and the tendency to use Western models. In the end, Zambia chose the policy of import substitution as a strategy for industrial development, despite its proven shortcomings in certain other developing countries outside Africa.

Zambia had no solid industrial base to build an import substitution strategy with. The mining industry was basically extractive. It had failed to provide for genuine backward and forward linkages with the other sectors of the economy and also provided no broadly based development. Furthermore, the mining pattern itself tended to confine itself to primary rather than to tertiary orientation. There is another aspect to mining activity to which the planners did not pay much attention. It relates to both the adverse environmental effects for the surrounding area and what was to replace mine activity as minerals are non-renewable resources. There could therefore have been a levy in the same way as a business enterprise provides for an amortization facility or fund.

The data in Table 3.4 show an increase in the average annual growth rates in the agriculture sector from 2.0 per cent to 3.2 per cent between 1970 to 1976, in contrast with continued food import up to 1975. The main cause of this was the annual percentage increase in the country's population, swollen by an influx of UDI refugees at a time when there was also a crop failure caused by draught.

The extent of progress made was illustrated by the range of food-stuffs displayed in some of the country's groceries and supermarkets. The availability of these products in remote rural areas did not seem to occupy serious attention. The programme should have stressed the need to transform subsistence agriculture farming methods. The continued availability of imported groundnut oil, canned mangoes, pineapples, meat and fish products, jams and jellies and cornflakes ought to have been a worrying factor to the country's development planners. The author is not implying a total substitution of imported food items but a substitution for those Zambia could provide comparatively cheaply.

Policy decisions to mechanise agriculture had considerable implications within the context of accelerating economic development. It was somehow short-sighted to have assumed outright mechanization of agriculture was the most effective to increase production. A major constraint to mechanization of agriculture in a developing country is the need to import technology, particularly machinery. Continued utilization of most modern agriculture implements requires a pool of reliable servicing facilities and spare parts. Naturally, this adds to costs which in many cases are beyond the means of the average farmer in the rural areas.

The extent to which these development plans contributed to the process of economic reform will be discussed in the last chapter. The achievement of the lofty objective of economic independence required more than the traditional instrument of a development plan. Above all else, it required the intensive and extensive mobilization of all human and natural resources and enlightened legislation.

The country's planning mechanism was at best, an exercise in identifying and curing hindrances to economic development. Identification of the main hindrances should not have been an end in itself. Some of the policies were designed to impart a dynamic thrust for structural transformation. Evidence in both the FNDP and the SNDP reveals a great measure of State capitalism underpinning the Zambian economy.

An exhaustive assessment of the impact created by each of the country's development plans would be rather time consuming. However, it is possible to have a bird's eye-view of some of the results. Keep in mind an assessment based on pure economic considerations rarely tallies with those based on social and cost benefits. And the implications of any plan, irrespective of whether it is bad or good tends to last beyond its own time span.

Additional evidence of the effects of development planning in Zambia like in many other independent African countries in recent years include revelations that they have devoted fairly large percentage of their total expenditures to programmes of education, health, nutrition, social welfare and social security. Most of these activities were formerly non-existent or were carried out by voluntary agencies, some of them subsidized by the respective governments. The high priority accorded to social development is reflected in the successive development plans. The cost of the social sector and general development is a balancing act between determining social benefits and social cost.

In a nutshell, social benefits can be interpreted to mean (*a*) gains in welfare which flow from a particular economic decision such as that reflected in a development plan; and (*b*) gains which go not to individual institutions or individuals but to the whole society. The concept of social cost, on the other hand, pertains to some activity or output by the society as a whole that need not be borne by the individual or firm carrying out that activity. For example, to what extent did planning activities pertaining to the mining industry, agriculture and import substitution reflect these social elements? If they did, then more evidence has yet to be established apart from the one pertaining to mining.

Wholesale criticism of attempts at planning the Zambian economy, particularly in the context of the FNDP has met with strong opposition in certain quarters. For instance, the Ministry of Development Planning and National Guidance opposed criticism but always had its own positive judgements to make. Supporters of the Ministry's views include Richard Jolly, an economist who contended that some of the criticisms were grossly misplaced, in particular when criticism showed a tendency to ignore the fact that a number of the planned targets were achieved ahead of time.[9] Jolly could very well be right, but he fails to acknowledge the fact that some successes were inspite of the provisions of the FNDP.

There were frequent instances of considerable hurry in trying to provide solutions. Some policy-makers' enthusiasm, coupled with some of the complications created by UDI, contributed to inter-ministerial conflicts and for several overlaps in area coverage. This is particularly noticeable in respect to operations of a number of state owned commercial corporations. There are several examples of lack of collaboration and coordination between the various quasi-public sectors in the SNDP document. At page 96, a statement can be found to the effect that an Indeco subsidiary was to establish eight units to machine-form and burn 160 million bricks while at page 244 in the same document is information about one Indeco subsidiary, having completed studies to establish large-scale mechanised factories to produce 160 million bricks.

There were other policies which created a state of confusion or conflict among the various ministries with regard to project execution mandates. For example, the Ministry of Finance on the one hand, and either the ministries of Mines and/or Commerce and Industry on the other, on matters of revenues and foreign exchange earnings or the Ministry of Finance and the Ministry of Development Planning and National Guidance (or the Ministry of Home Affairs) on the question of immigration policies and the implications of recruitment of skilled or unskilled labour, etc.

The Ministry of Finance is responsible for expanding the country's tax base in order to meet the growing budget requirements. The country's principal revenue source continues to be the mining industries. The Ministry of Mines and the Ministry of Industry tend towards a relaxation of taxes in order to attract investors in order to establish more industries, to accelerate import substitution and to create more job opportunities. The latter goal is also dear to the Ministry of Labour and to some extent the country's labour unions. However, conflicts arise in trying to accommodate the mining industry, which both ministries use as a base in their respective policy formulation. The Ministry of Finance recognises the mining output in projecting revenues and the Ministry of Mines also recognises this sector in attracting external investment funds for the expansion of the industry. Yet neither the FNDP nor the SNDP dealt with how such conflicts were to be resolved. A development plan has to be technically proficient. Had this been the case, there would not have been the need for so much reliance on the Cabinet to settle the issues which arose between ministries during the course of implementing the various plans.

Students of the development plans have also observed vagueness in delineating the framework and future role of cooperative societies. These institutions could have been integrated within the objectives of government's philosophy of humanism because of their compatibility. The attempt to organise the societies, especially in the rural areas, would help preserve Zambia's traditional way of life in a fast changing world. Cooperative societies should have received special treatment. They were supposed to ensure a preservation of mutual aid beyond individual self or immediate family interest(s). However, to be successful, they must incorporate the principles of inclusiveness and communal inter-dependence and reciprocity. Zambia's Cooperative Society establishment was pursued without guidelines to assure incorporation of these principles. Many of the institutional structures provided within the context of these plans failed to nurture indigenous entrepreneurship because of the same lack of clarity of their operational structure. Bruce Dinwiddy, a writer on African business, provides parallel insight with examples drawn from the Kenyan experience. Constraints which face African governments in trying to promote African enterprises and overall economic development seem to be the same. This is why he concludes by advocating the need for African governments to take a more enterprising attitude towards helping indigenous entrepreneurs.[10]

When the count is taken, as pointed out by Richard Jolly, one finds some degree of achievement. But the costs were too high in some places for any one to feel very satisfied. One reason for the failures was the lack of stress on private sector investment outside manufacturing. This situation is partly accounted for by the over reliance on the per capita growth rate ratio as a basic indicator in the analyses of development. Experience, not only in Zambia but also elsewhere shows that overall per capita growth is totally unsuitable indicator in a country with profound social and economic cleavages. Its major shortfall as a policy guide is that it fails to provide for a distinction between growth and development. Consequently, its generalized use as an indicator of progress is highly questionable.

The extent to which there has been any redistribution of income as a result of implementing the above plans during the 10 years following independence is the subject of the concluding chapters. We shall also answer questions on how attempts were made to lessen the export orientation and how the geographical constraints were handled. Meanwhile, suffice to say that logic dictates that development plans must be properly and clearly articulated in terms of a country's political, sociological and historical characteristics.

Those who planned the country's development also tended to leave certain vital details to chance. At the start of this analysis we referred to some of the prerequisites of a development plan. Assessment of Zambian development plans is bound to be similar to those of other developing countries. Perhaps it is worth recalling some other observed trends in this field, especially those pertaining to the economies of developing countries. In this connection, Zambia's development pattern seems to support economist Wassily Leontief's observation that there was an increasing tendency in developing countries to become increasingly reliant on the input-output models in planning and modernization of their economies.[11]

Had Zambia opted for a planned economy, all the basic planning functions would have been carried out by a centralized administration. This would have been an impossible task given the state of affairs at independence. The success of a planned economy is contigent upon the ability to dictate the total output of the economy. The central administrative body have to know the wishes of consumers and direct the expansion of each sector in a coordinated manner. The coordination of resource allocation would have been essential.

Had Zambia opted for a price system, resource allocation would have been based on the free movements of prices. In a price system a central government plays no major role outside of making legislation that ensures the smooth functioning of a free market economy. Zambia could not afford a price system given the polarity of her communities (Europeans or expatriates at the top and indigenous Zambians at the very bottom of the country's economic ladder).

On the other hand, the free market or what we defined as the price system appeared quite unsuited, given the prevailing conditions. Zambia inherited a colonial economic structure with several entrenched structures to perpetuate exploitation and repatriation of Zambia's wealth abroad. To have depended exclusively on the market system, as established in Zambia would not have produced development but would have perpetuated problems. To some extent this helps explain why the government found itself shouldering a greater proportion of the needed investment.

The Zambian experience confirms the view once expressed by the United Nations Economic Commission for Africa. That view is that many plans in Africa, not only in Zambia, are rendered inoperable rather early, and yet are kept in being through frequent, but unannounced modifications which seriously alter even their broad framework. The ECA further states that the problem is compounded by the fact that improvements in gathering planning data, particularly in relation to individual projects, compel frequent changes of course throughout a plan's life. To this end, it can be presumed that the value of many African plans as stable guides diminishes greatly before the end of their term.[12]

REFERENCES

1. Mwongozo, the TANU Guidelines passed early in 1971 by the National Executive Committee of the Tanzania African National Union, para. 28.

2. ECA, *A Survey of Economic Conditions in Africa, 1960-1964*, United Nations, 1968, p. 213.
3. International Labour Office, *Narrowing the Gaps: Planning for the Basic Needs and Productive Employment in Zambia*, Jobs and Skills Programme for Africa, Addis Ababa, January 1977, p. 7.
4. ECA, *Economic Conditions in Africa in Recent Years*, E/CN. 14/435, 31 December 1968, p. 169.
5. Ann Seidman, "Debt and the Development Options in Central Southern Africa: The Case of Zambia and Zimbabwe", *Recession in Africa* Jerker Carlsson (ed.), Scandinavian Institute of African Studies, Uppsala, 1983, p. 81.
6. Fergus MacPherson, *Kenneth Kaunda of Zambia: The Times and the Man*. Oxford University Press, Lusaka, 1974, p. 332.
7. Windsor K. Nkowani, "Planning Techniques in Zambia", *Economic Bulletin for Africa*, Vol. XII, No. 1, United Nations, 1976, pp. 67-92.
8. Richard Hall and Hugh Payman, *The Great Uhuru Railway*, Victor Gollanez Ltd., 1976, pp. 13-103.
9. Richard Jolly, "Zambia Six Years After: How Successful was the First National Development Plan?" *African Development Economic Survey*, Headly Brothers, London, pp. Z-10-11.
10. Bruce Dinwiddy, *Promoting African Enterprise*, Groom Helm in Association with the Overseas Development Institute, Lond, 1974, pp. 3-19.
11. Wassily Leontief, "The Structure of Development, *Scientific American*, 209, September 1963, pp. 149-164.
12. United Nations, *A Survey of Economic Condition in Africa, 1960-1964*, ECA, E/CN.14/401, p. 237.

Chapter Four

Integrating the National Economy

The pluralistic structure of almost all African countries, consisting of a traditional subsistence sector, and indigenous monetised sector and a foreign enclave, poses three main challenges: a small number of people generally found in urban centres, constitute the forces for modernization but they are surrounded by vast areas of rural backwardness where consequences of extremely low productivity are holding back development; foreign enclaves which governments have difficulty in controlling and integrating into national economies; and the fact that the participation of the indigenous population in the private sector of the economy is insignificant.[1]

The analysis in the preceding chapter helped to draw attention on the importance of development plans in a developing country like Zambia. That also reviewed the principle objectives, weaknesses and the strengths of the plans during the first decade of independence.

Integration of the Zambian national economy was one of the most difficult tasks after independence. This was especially true given the fact that the inherited colonial structure had been designed to serve foreign interests, (policy-makers were foreign and the executors were foreign). In other words, Zambia was 'foreign' in almost everything except that it was a territory inhabited largely by Africans, who had no responsibility for shaping their own destiny.[2] The underlying objective of the economic reform process that ensued was to bring about integration of the national economy. However, the pluralistic character of the country meant that the process of economic reform had to contend with physical, organizational, economic and social aspects.

Discrimination against indigenous Zambian natives (Northern Rhodesians) during both the colonial and Federation of Rhodesia and Nyasaland eras was an open secret. A policy of separate economic development or mild economic-apartheid, was already in force as part of the existing legal system. For instance, indigenous Zambians were only able to develop businesses in African townships on the pretext that these were areas where competition from expatriates had been prohibited. In practice however, effective control even in those areas, was still being exercised by the white minority through their overall control of the various municipal and city councils. Indigenous Africans were frowned upon if they ever dared to seek licences that would enable them to engage in business in areas or sectors pioneered by or reserved for expatriates.

It was against convention and the law of the land for indigenous Zambians to engage in commerce in areas designated for expatriate entrepreneurs. Society had been stratified into three quite distinct layers consisting of Europeans, Asians and Africans, similar to apartheid in the Republic of South Africa.

The shortcomings of colonial rule in denying the indigenous people trade and commercial opportunities were great. For instance, the educational structure was so devised that those who had gone through it found it extremely difficult to venture outside menial administrative job patterns. The system was devoid of any foundation on which to draw indigenous men and women capable of facing the cruel commercial world. Clerical officers, bricklayers, primary school teachers and office orderlies, to mention a few, tended to stick to their respective professions until death, except in retirement when some ventured into setting up village grocery stores. Their wages were kept deliberately below the poverty line so that they were incapable of accumulating sufficient capital to cut themselves off the colonial umbilical cord.

The above sombre picture of the Zambian economic structure was marginally better than in neighbouring pre-independent Angola with a white population in 1975 of 500,000. In Angola, any Portuguese irrespective of whether illiterate or poor could enjoy the fruits of good life. The only criteria was that he or she had to be white. Whites in Angola undertook such menial jobs as driving a taxi, tendering bars and hairdressing. On the other hand, the seven million black Angolans even after five centuries of Portuguese rule continued to be employed as servants, janitors or on plantations. Over and above, 98 per cent of blacks were illiterate and with no technical skills at independence.

This analysis will stress the need to redress such type of inequality of incomes and opportunities; imbalances between urban and rural areas in the distribution of development benefits, ownership of the main cogs of the economy,[3] and over-dependence on copper for the country's total revenue and economic wellbeing. All these issues had to be tackled by the indigenous population.[4]

PRE-INDEPENDENCE ECONOMIC IMBALANCE

A lot has been done in recent times to facilitate indigenization. All the same, it must be acknowledged that conditions in the two periods are markedly different. The pursuit of integration for the African businessmen into the general structure was hardest during the pre-independence era than has been the case in the post-independence period.

The background to the imbalance was, as stated earlier, created by the colonial administration. It was to be expected that changes were inevitable after independence. What follows is an attempt to show the role of certain indigenous Zambians in the field of commerce. Additional material will be provided so that the reader can make comparisons between the pre- and post-independence eras within Zambia and with other developing countries in Africa. Some of these efforts should be remembered because they spearheaded economic change. One such Zambian effort was the spate of boycotts organized by certain groups of Africans against discriminatory commercial activities. The boycotts were meant to push the system towards a more integrated commercial structure prior to the attainment of independence. It would have been safer for the organizers of those boycotts to have merely operated their own shops in their locations and kept out of the politically troubled waters.

There is a tendency in developing economies to concentrate every analysis exclusively on current economic activities, in total disregard of earlier periods. But, history must be set in its proper context and be broadly interpreted. Included in Zambia's history is the story of efforts undertaken at considerable odds by certain of its sons and daughters.

By the same token, expatriate groups guarded their economic privilege with absolute jealousy and as their birthrights. The only time when this was not insisted upon was when and where it became administratively impracticable for them to effect direct and immediate day-to-day control. There was an administrative laxity in the treatment of certain parts of the rural areas, designated as native reserves or native trustlands. But even in these few areas the non-indigenous group continued to influence affairs from their central vantage points. This was made possible through the established colonial administrative structure. The chain of command as it then existed ensured that indigenous person with even a little power were figureheads and not involved in the administrative structure. For instance, provincial administrators in Kasama took action against an African boycott of European business Thom and Company. That action clearly illustrated the ways which were used to give protection to European interests.[5]

Strong tribal rulers, like the Bemba Paramount Chief Chitimukulu and Paramount Chief Lewanika of the Lozi were unable to curb exploitive expatriate business activities in their respective areas. The power they could exercise were dependent upon the dictates of expatriate interests, whose influence could be brought to bear upon them with unforgettable consequences. In the final analysis, it was a prerogative of expatriate business interests that altered the course and direction of events in the tribal areas, in spite of the hollowed principle of *Indirect Rule*.

Indirect rule was a system of administration used by Britain in administrating its colonies and overseas possessions. The system was favoured by Britain in administrating its colonies and overseas possessions. The system was favoured because Britain economised its administrative expenditure and use of British human resources by using or ruling as much as possible through African chiefs. Apart from the insufficiency of British administrators, the policy of indirect rule was considered as likely to be more acceptable if foreign rule could be exercised through the respective chiefs. Protagonists of indirect rule like captain (later, Lord) Frederick Lugard in Nigeria stressed the need to educate and guide the native chiefs to understand the problems of modern government and leave them to explain the ideas to their people. However, those who followed after Lugard found in this a better way to oppress the natives economically and commercially.

The colonial administration failed to spearhead an integrated approach in the field of economic development. This was most obvious in the agriculture sector where administrators followed a mixture of piecemeal plans. As a result, commercial agriculture was dominated by non-Zambians, assisted by the administration. Official support of the administration to expatriate farmers included easier access to credit facilities, which helped assure their success. In addition, feeder roads were constructed in places that made expatriate farming areas more attractive and commercially viable. Tracks of land were cleared at government expense, as a subsidy to expatriate farmers, in such areas as the Mkushi Block, Kabwe and Lusaka. These areas were cleared to encourage the cultivation of tobacco and maize cash crops. Similar help was extended to expatriate cattle farmers, engaged in ranching or dairy production in parts of the country. The mechanism was such that only the laziest among expatriate farmers could fail.

Over and above all this was the creation of highly subsidized country-wide marketing and distribution institutions to collect the produce offered for sale by the non-African farmers. One such institution was the Grain Marketing Board (GMB) in both Northern and Southern Rhodesia. The Board constituted an effective mechanism for providing comfort to expatriate farmers without the administration being seen as actively engaged in favouring one sector of the community against the indigenous majority in the population.

Measures initiated in favour of African farmers were to a large extent weak and represented no well-thought-out policy. The indifference of the colonial administration to the plight of the indigenous farmers, in particular the subsistence sector, was part of a mild economic-apartheid.

Colonial administration attempts to correct the deep-rooted and glaring imbalances were the setting up of the Ministry of Agriculture and the Department of Veterinary and Tsetse Control. It would be a misrepresentation of facts to attach much importance to either of these administrative machineries, as there is very little evidence in support of their getting anything done. Missing elements in both policy and objectives were the intention to assist indigenous African farmers to obtain control of the agriculture sector or even become equal partners in development. Whatever benefits accrued to the African farmers were accidental. The same also applies to such other colonial initiatives as extension schemes for the indigenous tobacco farmers surrounding the European Mkushi Block. The facilities extended through the schemes were not only insufficient by markedly inferior in comparison to what had been extended to expatriate counterparts. One of the undeclared reasons behind the setting up of the African extension schemes was to attract a pool of cheap African labour for the nearby European tobacco farms. The white farmers did not have to provide such expensive infrastructure as accommodation for extra farm labourers during the harvest periods.

Along with the creation of the above government ministry was the establishment of the Agricultural Rural Marketing Board (ARMB). This institution was expected to help the indigenous farmers to obtain simple farming implements, such as: ox-drawn ploughs, ox-drawn (scotch) carts or wagons, hoes and hand-axes. The institution was also required to purchase agricultural produce from African farmers. The duplicity of intention by the colonial administration and the vested interests of the non-African farmers, who controlled the Grain Marketing Board, combined to negate the supposed effectiveness of the ARMB from the African farmers' standpoint. This duplicity of the colonial structure could also be seen in the way the GMB exerted its control on the price structure of the African produce purchased by the ARMB. Prices offered for comparable African produce were statutorily required to be lower than those offered by the GMB.

Non-African farmers abused their privileges with the blessings of such bodies as the GMB. Unscrupulous white farmers bought African agricultural produce at low prices and resold them at higher prices through their privileged institution, the GMB. The legal structure had superimposed the GMB on the ARMB. The same legal framework encouraged the plunder and exploitation of the indigenous farmers by the white farmers. Provisions under Sections 3 and 4 of Chapter 246 of the Act which established the GMB, buying agents and buyers of African-controlled produce were empowered to make certain deductions from the final price paid for products purchased from an African producer.[6]

Maybe such action was considered necessary in order for the buying agent to defray his or her transportation and storage costs (assuming this was applicable?). However, the absence of criteria for determining the scale of such deductible costs meant that many unscrupulous white farmers found in the system another haven for continued exploitation of the African producers of many cash crops. Secondly, the architects of the legal instrument itself failed to appreciate the fact that it would have been judicious to have absorbed those costs in the form of subsidies in order to encourage the African farmers to become commercially viable.

The consequence of this was that the GMB, a buying agent, or a direct consumer (normally non-African) of such items could make any such deductions in respect to:
  (i) any levy imposed under the provisions of Section 15 of the African Farming Improvement Fund Ordinance;
  (ii) transport charges; and
  (iii) handling charges.

Hence African farmers neighbouring European commercial farmers sold their maize produce at prices of say K5 per 100lb. bag. Maize produced on the African farms even if produced commercially was regarded or classed together with that produced and sold by the subsistence farmers in the remote rural areas. Its grade, even if comparable or superior to that of the European farmer(s) was of no consequence so long as it was being traded by an African. Therefore the buying agent offered less than K5 per bag (20 per cent to 30 per cent less). In the meanwhile, the buying non-African agent incurred no transport or storage costs because these facilities were already provided by the Government through the GMB.

The indigenous farmers, and particularly those who sold their produce to any of the privileged groups referred to above, could never hope to earn enough to become independent of continued exploitation. The structure was such that the former would continue to suffer exploitation at the hands of the expatriate farmers, no matter how hard the land was worked.

PRE-INDEPENDENCE INDIGENOUS BUSINESS

For those who have never physically experienced the adverse effects of direct colonial economic oppression and exploitation what follows may seem irrelevant or academic. This analysis of pre-independence business is an attempt to put early indigenous pre-independence business entrepreneurs in the right perspective. It also provides valuable information on certain organizational structures in terms of both the origin and background to the beginnings of entrepreneurial zeal.

The modern Zambian African should identify himself with his past. There is therefore in this chapter a short historical account of past Zambian economic acumen. There is also an account of experiences which indicate that those who have profited from previous economic vantage positions did not give up without resistance and clandestine sabotage.

Native or indigenous commercial acumen during the colonial period was confined to few areas and disciplines. It concentrated in the rural areas covering transport, ownership of retail shops and groceries and eating places or restaurants. Late additions in this field were the rural hotels, bars, private law practice firms and free-lance journalism. Against this background, the analysis will review not only some of the elements but also the importance of both the Mulungushi and the Matero economic reform programmes. Also taken into account will be reform programme's timing in light of the objectives enunciated and their relation to the process of integrating the national economy.

Perhaps the most active area of rural commercial activity was in the Luapula Province with a little spread into the neighbouring Northern Province. Commercial activity by the indigenous population on the line-of-rail and in the copperbelt were peripheral in nature, except in the last days of colonial rule. The majority of indigenous Zambian business enterprises expended their resources and energies in activities related to simple commerce and transport. The early group of businessmen in the Luapula area were concentrated around Mansa (formerly Fort Rosebery), Samfya, the villages of Lukwesa and Mwansabombwe. Prominent indigenous businessmen included the Musango Brothers, the Kalyafye Brothers, the Mwenso Brothers, and the Chisakula Brothers, Nsemukila, Kosta, Luka Mumba, Kapesha, and Kashimbaya. In the Northern Province on the other hand, mostly around Mbala and Mpulungu, were men like Wilombe, Shicinsambo and Chinunda in Kasama.[7]

The other parts of the country had smaller businesses but a similarity in the pattern of operations. In the Western Province (formerly Barotse Province), the list of businessmen included people like Willie Harrington and Mubonda and Sons. In the Southern Province it was people like the Dimba Brothers of Chikankata and Zaloumis of Choma. In the Central Province there was Thom Manda and Sons, whose fleet of bases plied as far afield as Mumbwa and the Luano Valley from their central base at Mkushi River. Thom Manda and Sons owned a chain of stores specializing in the sell of hardware goods and other consumer durables. In the Eastern Province Ziloli Mawere owned a chain of stores and buses similar in character to that of Thom Manda and Sons in the Central Province.[8]

A common feature of the pioneering indigenous Zambian business entrepreneur in the Luapula, Northern and Copperbelt Provinces was their common background. Almost to the man, they had all spent the early part of their youth in Shaba Province in Zaire (formerly Katanga Province of the then Belgian Congo). Many of them had no formal education. However, they had acquired business experience as salesmen or storekeepers working in Greek firms in the Congo. Their starting capital was largely out of individual savings. But there was no help from the nation's financial and banking institutions to enterprising indigenous businessmen during the colonial era. Their survival instinct, gained from the Greek traders, was comparable to that displayed by the Asian business community.

Another common characteristic among the business pioneers of the Luapula Province was their bias for trade in fish caught from the River Luapula. The Mwenso Brothers, Kalyafye Brothers, Nsemukila, Kosta, Luka Mumba, Kapesha, Kashimbaya—all at one time or another had been fishmongers. The fish trade had very few overhead costs and was thus a potentially lucrative line of business. This was clearly demonstrated by the Greek businessmen living in Zaire who owned fishing vessels on the Luapula River. The demand for fish in Zambia, especially in Luapula, the Copperbelt and Northern provinces was partly a result of the shortage and high price of meat. Fish from these areas was also in high demand as far afield as Southern Rhodesia for the low income group, mostly Africans whose monthly wages made the price of meat beyond their reach.

With gained experience on the basis of the above, many African traders branched out at a later stage into trading in second-hand garments purchased from the Zaire towns of Sakania and Mikambo. These were sold by hawkers, mainly using bicycles. This kind of trade was nicknamed "Kombo". A corruption of the place of origin "Mikambo"[9], Kasumbalesa and other Zairean towns were also places where these items were purchased. The emergence of this trade was dictated by what economists like to call the forces of supply and demand, coupled by rise in income along the line-of-rail and the Copperbelt. However, demand-pull was also in effect. Increased purchasing power in the Luapula and Northern provinces came through remittances being sent into the rural areas by miners on the Copperbelt for the upkeep of their families left behind. This pattern is similar to financial benefits derived by Malawi (before 1975) and Lesotho, who sent migrant labour to the South African mines.

This background also helped set the pattern of subsequent development. For instance, the emergence of a network of indigenously operated transport services, connecting most rural areas and the rail-line. The Musango Brothers, Mwenso Brothers, Thom Manda and Sons, etc. were working in an area where the colonial administrator did not give priority.

There was little Zambian business in the more sophisticated fields, especially establishing of hotels and restaurants. The most notable pioneering venture was the Mansa Hotel established by Luka Mumba. This was in addition to his shops and a boat service at Samfya. He had gained enough confidence and experience as an entrepreneur and this new line of commercial activity came as no surprise. Other hotels of a relatively low standard were later established at places like Mwansabombwe and Lukwesa by the Chisakula Brothers. At about the same time, there were similar hotel and restaurant activities by women like Anna Chilombo in Mansa and Anna Lengalenga around Katuba near Lusaka. Africans were generally discriminated against in commercial activities by expatriates. The entry of indigenous women was perhaps harder still. They had to overcome the resentment of the African menfolk who by tradition always gave the womenfolk a secondary role in society, except in the kitchen.

These activities contributed to pioneering and influencing indigenization of the economic structure. The significance of the above activities must be weighed against the difficult circumstances in which they were nurtured. For instance, there was total lack of government financial support facilities to indigenous businessmen. It, therefore, became a matter of survival of the fittest in an unaccommodating economic infrastructure. Existing commercial and financial institutions like Barclays Bank or Standard Bank, branches of multinationals, made no attempt to facilitate the entry of African businessmen in the country's economic fabric.

To some extent, these are some of the activities which contributed to the sharpening of legislation on indigenization in Zambia. But the adoption of a policy of indigenization and in particular integrating the indigenes was not until the 1968 reform process was set afoot.

Integrating the economy could only have come about with the government helping the indigenous people penetrating other sectors of the economy. However, the spread of indigenous business entrepreneurship into such fields as journalism had been without much direct government support. It was nonetheless initiated by Titus B. Mukupo and Tanner Malinki, with the establishment of the first-ever indigenous free-lance journalistic bureau. As in the case of the African early lawyers,* the journalists located themselves on Lusaka's prestigeous Cairo Road. These initiatives demonstrated above everything else, the fact that the trend was beginning to take shape and that the country's economic scene would witness the dismantling of the barriers which kept Europeans and expatriates in general at the top of the economic ladder with Africans relegated to the bottom of it. Even the area of smaller consumer activities had not been spared foreign domination. Asians of Indian origin, assisted largely by loans granted by expatriate banks operating in the country had managed to obtain a sizeable share of petty consumer trading. Some Asians had also established a virtual monopoly of wholesale trading for various merchandise.

It is therefore the aim of the subsequent section to examine not only the overall policy of integrating the economy but also how the strategy was laid for its implementation. Integrating the economy meant bringing into the mainstream those sectors of the economy which had been denied the opportunity. Within the Zambian context as in Ghana, the problem was how to raise the socio-psychological status of the citizens vis-a-vis the aliens to positions hitherto reserved for foreigners.

INTEGRATING POLITICAL INDEPENDENCE WITH DEVELOPMENT

The task of integrating political independence with development

---

* Edward J. Shamwana had established his law firm in Cairo Road of Lusaka in a place previously owned by his predecessor, Fitzpatrick Chuula.

given the background of colonialism was made for difficult because of the dualistic nature of the economy. An analysis of dualism of the African economies by the United Nations in 1958 stretches as far as the beginning of the twentieth century.[10] Dualism of the African economies is characterized by the enlargement of the money economy. In other words, it involves the shift of resources from subsistence production to production for sale. On the other hand and parallel with the above is the subsistence sector which is characterized by four features. They include (*a*) a lack of significant specialization; (*b*) a lack of regular production surplus for sale; (*c*) use of rudimentary or primitive technology in production; and (*d*) domination by the indigenous population whose main use is the supply of cheap labour for the other sector (the money economy).

The attainment of political independence was an important step in the effort to integrate the Zambian economy. Moving from political independence to economic development was a logical step in the country's history. The old rules governing the country's economic game had to be reformulated. Needed were workable solutions capable of quickly integrating indigenous Zambians into commerce, agriculture and other economic sectors, including the State government itself. Some indigenous people had already set their eyes on certain targets while the non-indigenous Zambians had their misgivings about the capability of the new political masters.

Zambia acquired the political means for correcting the inherited economic imbalance with the attainment of political independence. However, the country's new rulers soon realized that implementing economic changes was an uphill battle. First, they felt the need to continue attracting foreign capital and expertize. Second, they desired to retain what was available without seeming to back-track on the pre-independence manifestoes to indigenise the economy. At the same time, giving a sense of security to the original investors, especially the mining companies and several commerical farmers.

The Zambian economy started off with a relatively impressive bargaining stance because of its mineral wealth and a surplus of foreign reserves. It had repudiated the Royalty Agreement on the eve of independence on its own terms. It had pulled off the first-ever successful economic challenge in a very highly protected area i.e., the mining sector. In that move, the would-be Zambian government called for the abolition of the remaining life of royalty payment on mineral concessions. The BSA company was offered and paid two million pounds sterling on a take-it-or-leave-it basis. In other words, the new African Government had refused to pay the company despite the concurrence of the British government.

It was therefore to be hoped that savings therefrom would be used towards integrating the economy, especially in support of the hitherto neglected facets of the subsistence sector.

This act had been a good test case in as far as the future patterns of possible acquisition and control of certain sectors of the country's economy and in integrating others still dominated by expatriates. However, the role of the mining sector in the Zambian sector should be set in its right perspective. At that time, the action did not distract from the government's concern with revenue returns and not upsetting the applecart. In particular, the government felt the need not to offend those parties on whom the country depended for continued viability of the mine industry. The dilemma of economic vulnerability has always been a major constraint for the Zambian government when it came to deciding on certain required actions. The Government had grown to fear rather than respect the controlling power of the BSA Company, whose principal shareholder was the Anglo-American Corporation (AAC). Everyone in government circles realised the need to tread with caution in the short, medium- and long-term interest of the country. Zambia's brinkmanship involved the dilemma of integrating the national economy and recognising its own limitations. Short-term solutions needed to be reconciled with long-term effects.

This awareness on the part of the new Zambian government was well founded. First, both the AAC and RST were recognised as vital in determining how the mining industry would run. Secondly, both companies had the external means and power, if so desired, to cause harm to Zambia's short- and long-term economic development. The nature and structure of the mining sector in Zambia indicated that the country was destined to continue to depend on the expatriate mining companies· for necessary expertise and external funds for further investment. This was also true for other development projects. Accordingly, the rules of the game could only be changed by tactful negotiations. The Bemba proverb *Mukolwe pa kukula epo afume ipindo* (translated, means that it is at youth that a cockere may break its wing) is a fair summary of the dilemma that faced Zambia in challenging the supremacy of the giant expatriate mining companies.

Without the abolition of the concession rights, the BSA Company could have treated the new government as any other prospector. This is because the Laws of Northern Rhodesia, at Sections 19 and 20 of Chapter 91, defined the rights and obligations of prospectors vis-a-vis the concession-holders. It was mandatory under the law for every prospector to obtain permission of the property owner (in this case AAC and RST) before prospecting. After independence, the new major prospector was the government. The concession owners could have inhibited the new government from carrying out its work towards an integrated development programme which it felt was demanded by the new political setting. It is possible that concession owners could have been moved by good common sense and mutuality of interest to allow the government to exercise its sovereign right without seeming to insist on its pound of flesh.

It was fortunate that the loss of face suffered by both the BSA Company and its principal shareholders (Anglo-American Corporation) over the royalties settlement did not result in any serious adverse economic effects for the new Zambian Government and the economy as a whole. The royalties action did not alienate a considerable number of outsiders and the expatriate miners who kept mines functioning smoothly did not take sides with management. The importance of this is accentuated by the fact that the colonial labour laws had prohibited the apprenticing of African mine workers on-the-job. Thus, there was a serious lack of skills for quick replacement of expatriates.

To see whether this step was a means to grafting good economic sense with political expediency it is necessary to review the background to UNIP's pre-independence constitution. To a large extent, the Party's constitution was a reflection of the letter and spirit contained in several manifestoes adopted before and after independence. Among the issues and solutions related to integrating the economy were:

(*i*) the abolition of all forms of discrimination and segregation based on race, colour, tribe, clean, and creed in commerce;

(*ii*) the need to secure the most equitable production and distribution of the wealth of the country in the best interest of the people;

(*iii*) protecting and promoting trade, industry, and agriculture in the people's interest, through legislation; and

(*iv*) cooperating with industrial and other organizations which complied with Party's policy and principles.

Some of the difficulties of this agenda are similar to those in Kenya, Nigeria, Tanzania or Ghana. To appreciate the enormous tasks involved in economic development one has to consider the socio-economic policies and structure imposed during the colonial period, as Cadman Atta Mills tried to show in case of the Ghana.[11] Expatriate interests controlled and owned the country's largest business enterprises. Any of the budding indigenous businessmen were overshadowed immediately after independence by the intrusion of aliens and expatriates who opted to acquire immediate Zambian citizenship. Some actions of these new citizens in the aftermath of independence merely served to strengthen and perpetuate the continued exclusion of the indigenous population from the country's economy. The results were the opposite of what the UNIP had hoped for. This was a dilemma which had to be challenged and overcome. (Chapter 5 will describe the problem and solutions in detail.)

These development problems were the subject of the frequent pre-independence speeches at several political rallies, especially those held at Mulungushi Rock.[12] Apart from rhetoric, there was no mistaking the desire of the future Government to take over control, or at least have the major say in the economy. These issues and others will be analysed in greater detail particularly in Chapters Five and Six as well as in the conclusion to Chapter Seven.

Suffice to recall that the principle of laissez-faire was the order of the day especially where expatriate businesses were concerned. The country's mining sector from the discovery of copper on the copperbelt and lead and zinc at Kabwe (formerly Broken Hill) was not subject to direct government interference. Avoiding direct intervention by the state during the colonial era allowed for the industry's low profile, in spite of the fact that mining was the most important economic sector in the country. Nurturing the mining sector under those conditions placed the former administration responsible for creating, abetting and perpetuating artificial restrictions which hindered the rapid advancement and integration of African workers in the mine industry and their skills. Integrating the mining sector into the economy beyond its assumed role of generating income to pay for limited development was another major preoccupation of the government.

In the absence of a coherent national policy on economic indigenization, the expatriate and the newly enfranchised business interests found it relatively easy to exploit Zambia's post-independence economic boom. Many of the profitable expatriate business interests did not re-invest in the Zambian economy or help in its future development. They chose to repatriate their profits.

Government concern about this trend was probably the background to the President's statement made in 1968, when he addressed a group of British businessmen on profiteering by expatriate business enterprises. The Zambian economy had grown quite strong in the four years after independence and net profits accruing to some firms amounted to between 25 and 30 per cent of turnover. Businesses were able to declare dividends between 100 and 150 per cent of capital. Even for the few who had suffered losses, their parent companies were able to enjoy administration fees as high as 10 per cent.[13] The President's statement was intended to reassure others to consider investing in the Zambian economy, which had proven itself resilient despite the difficulties created by the UDI.

These observations provide a background to understanding and appreciating the motivations for actions taken by the Government during the first decade of independence. Perhaps the most difficult balancing act for the government was how to set the timing for its actions. In other words, the Government had to:

(i) determine who was to own the means of producing the country's wealth, notwithstanding its own declaration in favour of a mixed economy in regards to areas: of government monopoly, of shared government to government participation, where the above intermingled with limited private enterprise, where the government could participate with private enterprise, and by mutual agreement, the government could take over full

control at a later stage, for cooperation with or without government participation and for the indigenous population;
   (*ii*) establish the means of production, control and ownership;
   (*iii*) set production priorities, in view of existing constraints imposed by resource availability and an underdeveloped infrastructure; and
   (*iv*) maintain control over distribution responsibilities without scaring away the much needed inflow of capital.

What is clear from the above is a picture which shows that there were sectors in the economy which were more amenable than others to integrating into the economy. In a sense, the findings suggest support for the view that certain sectors of the economy such as mining were more difficult to integrate, but more rewarding in terms of its impact on total output. The country's economic structures only made very slight shifts from those inherited a decade earlier. Dependence on non-African markets for both exports and imports were not reduced to any substantial degree. The development of trade with neighbouring countries or intra-African trade with those not in the South was still very rudimentary. However, the Government had begun to take steps to correct a number of these problems, as will be noted in the next chapters.
The state had not been totally successful by integrating the economy at the end of the first ten years of independence. Its intentions of preventing foreigners from continuing to dominate the ownership of production and distribution channels in order to reduce the vulnerability of the security of the state was a task which could not be achieved in so short a period. Consequently, the policy of indigenization which was heralded by Mulungushi and Matero economic reform programmes, described in greater detail in the subsequent chapters provide some of the answers in this area.
Many elements combined to undermine the possibility of success in integrating the economy, especially through the policy of indigenization. There was a certain degree of administration fumbling at the government level itself. This pattern has seemed to plague many African countries—Ghana, Kenya, Nigeria, etc. Added to this, were activities of aliens who sought and managed to

evade these measures through ownership fronts, exceptions to the rules and citizen changes. The sum total of it all was that local citizens got cheated in the transfer process because of the lack of established government machinery for the purpose of protecting indigenous businessmen.

The difficulties encountered in integrating the post-independence economy in Zambia was typical of any African state which had inherited a colonial economy. According to Ann Seidman, these countries end up stressing the expansion of social and economic infrastructure and the associated administrative bureaucracy (with or without marginal changes in the institutions). Seidman goes on to state that "to finance the resulting increased expenditures [they] may borrow funds and seek to attract transnational corporate investment to expand exports in the hope of augmenting future tax revenues and foreign exchange earnings".[14]

REFERENCES

1. ECA, *Africa's Strategy for Development in the 1970s*, United Nations, pp. 2-3.
2. Government of Zambia, "Zambia 1964-1974: Ten Years of Achievement", Zambia Information Services, Lusaka, 1974.
3. Jonathan H. Chileshe, "Zambia" *Indigenization of African Economies*, Adebayo Adedeji (ed.), Hutchinson University Library for Africa, 1981, pp. 103-106.
4. William Tordoff (ed.), *Politics in Zambia*, Manchester University Press, 1974, pp. 15-19.
5. Kapasa Makasa, *March to Political Freedom*, Heinemann Educational Books, Nairobi, 1981, p. 48.
6. *Laws of Northern Rhodesia*, Vol. VII, 246 Section 3 and 4, pp. 2-3; *Federal Act No. 23 of 1957* as established as the Grain Marketing Act in 1957 by the Federal Legislative.
7. Jonathan H. Chileshe, *op. cit.*, p. 82.
8. Based on author's personal interviews with Michael C. Sata and Alderman Safeli H. Chileshe (Luapula and Northern Provinces), Reuben C. Kamanga (Eastern Province), Alderman Safeli Chileshe and Henry Meebelo (Western Province).
9. Mikambo is a name of a Zaire town on the border with Zambia (almost adjacent to Mufulira). However, the name was adopted to also describe the second-hand merchandise obtained from the Shaba (formerly Katanga Province) towns of Mikambo, Sakania and Kisenga and sold in the then Northern and Southern Rhodesia (Zambia and Zimbabwe).
10. United Nations, *Structure and Growth of Selected African Economics*, (New York, Department of Economic Affairs, 1958) pp. 1-5; Rhodesia, "RACE" Vol. IV, No. 1, November 1962, pp. 73-87.

11. Cadman Atta Mills, "Dependent Industrialization and Income Distribution in Ghana", *Industrialization and Income Distribution in Africa*, Codesria Book Series, 1980, pp. 61-62.
12. Mulungushi Rock lies on the Kabwe (formerly Broken Hill)—Kapiri Mposhi Road on the eastern banks of River Lunsemfwa. It is this place where the strategic plans for launching and carrying the United National Independence Party's (UNIP) independence struggle were drawn up.
13. K.D. Kaunda, "A Plea for Understanding", address to the Overseas Development Institute, London, 18 July 1968.
14. Ann Seidman, "Debt and Development Options in Central Southern Africa: The Case of Zambia and Zimbabwe", *Recession in Africa*, Jerker Carlsson (ed.), Scandinavian Institute of African Studies, Uppsala, 1983, p. 90.

Chapter Five

The Mulungushi* and Matero** Reforms

Our basic problem has been to try and undo the mistakes of the past and to give correct emphasis on the real meaning of the welfare of the people of this country.[1]

The overall review given in the preceding chapters has exposed the openness and vulnerability of the Zambian economy both externally and internally and that the government had little or no control over it in the early years of independence. The dual nature of the economy was clearly visible in the early years of independence. The pattern of production was also narrowly confined in the urban sector along the rail line between Livingstone and the Copperbelt, was based on modern technology and was directed towards the money market. On the other hand, there was the rural sector in which lived the bulk of the population with a production pattern that relied predominantly on crude technology directed at meeting subsistence needs. Production of most primary products with any export potential continued to be carried out, only with minor modifications, in the traditional land tenure system.

It was therefore imperative for the Government to seek appropriate ways and means to undo the mistakes of the past and to give needed future direction. This had to be done by income redistribution through various tax incentives and redirection of industrial programmes. This was apart from bringing about a balanced sectoral and provincial economic development.

During the period under consideration, revenues derived from the copper industry constituted the only base for building the

---

* Mulungushi in here refers to the site of the Mulungushi Rock, situated off the Great North Road half-way between the towns of Kabwe (formerly Broken Hill) and Kapiri Mposhi. It acquired Zambian and international fame by virtue of being a birth-place of the United National Independence Party (UNIP). Subsequently, because other important or watershed speeches have been made from the same spot, especially the 1968 Economic Reform Programme.

** Matero is the name of a township or a western suburb of Lusaka.

country's economy. High copper prices underpinned most of the recorded successes under the FNDP. Similarly, weakening of copper on the London Metal Exchange (LME) during the major part of the SNDP cast a shadow of gloom over many sectors of the Zambian economy. What this development underlined was the fact that there was an urgent need to develop foreign exchange alternatives as part of a strategy of self-reliance.

There have been so many watersheds in Zambia's history that it is quite a task to single out any one event as being the most prominent. Some of the early steps to give the new Zambian state its own identity included stopping denominating the national currency in "pounds" but in "Kwacha". The economic package introduced under the Mulungushi Economic Reform Programme (MERP) in 1968 is one of these and will be highlighted as indicative of the country's concern about conditions in the commercial, industrial and financial sectors of the economy.

It is possible to view the MERP and the subsequent Matero Economic Reform Programme (1969) as background to the preceding analysis on economic integration. They were designed to foster economic growth and help achieve economic development. It is important to preface the Mulungushi discussion with the following summary of two theories:

(*i*) The growth theory states that within a certain time-frame there are certain determinants of the rate at which an economy will grow. The theory assumes an aggregation of national income, consumption, capital and total employment. The fact that this theory is highly abstract and biased towards mathematical analysis rather than anything else is beside the point. What concerns us here is to draw a distinction.

(*ii*) The economic development theory proposes fundamental changes in the structure of the economies of developing countries being reflected in total growth and growth in *per capita* income.[2] The main signposts according to this theory in addition to what is stated in footnote 2 are changing and increasing the importance of industrialization as opposed to agriculture, reversing the trend of rural depopulation, lessening dependence on imports for more advanced consumer and producer goods, lessening dependence on primary agricultural and mineral commodities and diminishing reliance on foreign technical assistance and aid.[3] The main object is, as earlier stated, to raise the standard of living or welfare of the people.

THE MULUNGUSHI

In April 1968, at Mulungushi Rock, the United Nations Independence Party (UNIP) set about to remove foreign domination of the Zambian economy and economic life. It acquired control of most major means of production and services. These actions were also aimed at establishing a firm foundation for the development of genuine Zambian business.

It will become clearer in the subsequent parts of this chapter that most of the elements contained in both the Mulungushi and Matero economic programmes had a great measure of futuristic appeal. However, it was not until the United National Independence Party's (UNIP) National Council in November 1970 that the nation was able to obtain a full elaboration of those reforms. It emerged that the State intended to assume a major say over the economy by participating actively in all key economic areas. It is doubtful whether the mechanisms were fashioned after either of the two theories earlier summarised. Similarly, there was little evidence that they were fashioned after the Rostovian, or Western model of economic growth or the Latin American (Raul Prebisch's centre-periphery) model. However, there are several elements to be discussed which border on most of what these theories try to demonstrate.

Before Mulungushi

A serious gap in effective control of the economy by the Government, at all levels, was only one legacy of colonialism. Furthermore, there was an almost total absence of indigenous leadership in most expatriate firms. There were almost no indigenous Zambians serving on the Boards of Directors in the industrial and commercial sector especially banks (Barclays, Standard, National and Grindlays), building societies, insurance firms, multinational corporations (e.g. Lonrho complex, British-American Tobacco) and the mining companies (e.g. Anglo-American Corporation, Roan Selection Trust). However, RST had already taken the initiative by appointing Alderman Saféli H. Chileshe to some of its directorships.[4] The government of course turned its attention in that direction.

Only at a much later time, in the aftermath of Mulungushi, did changes begin to appear. Most significant was the appointment of indigenous Zambians as executive chairmen of the Standard Bank of Zambia Limited, (Elias M. Chipimo) and Barclays Bank of Zambia (Africa B. Munyama). Admittedly, most of the then available indigenous talent was employed as government civil servants and could not be spared for the private sector. However, the basic reason for lack of action on the part of the private sector was the high degree of vested interest associated with prevailing conditions south of the Zambian border where many expatriate companies operating in Zambia were headquartered.

In the few cases where some attempt at indigenization or localization of personnel had been attempted, the motivating factors were almost always the same, fear and/or self-interest. Indigenization provided an insurance against possible nationalization by the State. Cases of genuine desire in assisting the government in carrying out its declared policy through localization were few. However, the new development expressed by the government through the MERP provided an ideal opportunity to those who had hitherto been unable to carry out such a policy. They were able to introduce modifications to their operational structure which permitted for an element of localization of personnel.

Naturally, the expatriate firms chose the easiest way whenever they embarked upon localization. Localization of the post of personnel officer in their respective companies seemed easiest. Opinions may differ, but the offer of the post was not intended to confer any real authority to the incumbent Zambian. In general, the new Zambian personnel officer became a mere post office, a showpiece intended to hoodwink the government and to protect expatriate management. The practice served the expatriate companies well since the Zambian acted as a buffer against any storm coming from his fellow Zambian workers down the line and the government at the other end of the spectrum. It was also common practice not to appoint a Zambian as personnel officer in any company where there was still a predominance of expatriate workers.[5]

The Industrial Development Corporation (Indeco) and the machinery that Indeco applied in localization, as we show next, shows the vicious colonial economic cycles which the Mulungushi

Programme found itself dealing with.

Mulungushi and the Industrial Development Corporation

The Industrial Development Corporation (Indeco) was not an offshoot of the Mulungushi Economic Reform Programme. However, the latter used the former in pursuance of its objectives. Indeco was created as a Loans Board in 1951 by the government of Northern Rhodesia, basically to help struggling private expatriate entrepreneurs in establishing businesses. The original Indeco structure confirmed the already discussed colonial machination of ensuring the presence of Europeans at the top of the country's economic ladder. Indeco eventually became a limited liability company of private investors (expatriates) which helped create the economic polarization of the country's communities. Indeco financed only expatriate business interest and those in which the corporation itself had minimal direct indigenous Zambian participation. This situation continued until the government bought all the shares in 1964. Thereafter, the structure became a state-run agency.

Indeco's expanded activities immediately after independence were a direct result of direct government involvement in industrial development. In particular, the government used the corporation to effect the programme enunciated within the Mulungushi Economic Reform Programme. Thus, Indeco, like the Ghana Industrial Holding Company (GIHOC), had to promote, finance and manage several industries on behalf of the government. Indeco was entrusted with the responsibility for implementing the country's new policy which was intended to accelerate the creation of both small and large industries.

Indeco's operational structure was changed in 1968 with the acquisition by the State of a major controlling share in 24 companies. The companies were involved in brewing, transport, building supplies, finance, hotels, chemicals and property. The main task of the corporation then turned out to be welding those companies into a coherent group and ensuring that each fell into step with official policy regarding industrialization. Subsequently, Indeco became a subsidiary of the Zambia Industrial and Mining Corporation (Zimco) in 1969.

The first Mulungushi economic reforms, dated 19 April 1968 transformed Indeco into a very large conglomerate. The old structure was therefore beset with the problems of acquiring new commercial and industrial interests and reconciling itself to a totally new economic, social and political environment. This transition must have been foreseen, as shown in a policy statement by the Minister of Industry and Mines, E.A. Kashita, when he stated that "no economic, or industrial plan could mature outside the political and other factors and certainly no investor (other than the Government itself) would risk his money until political, fiscal and other considerations are revealed."[6]

The existence of Indeco was an advantage to the Government at the time of initiating the Mulungushi economic programme. The government turned to Indeco to negotiate on its behalf with the private companies in which the State had declared an interest. Indeco was subsequently instructed by the State to refuse any payment for such things as business goodwill or for future profits to any of the expatriate companies so acquired.[7] The dilemma in this regard was the likely effect on those expatriate businesses, which had been spared the axe in the first round. Would they close down? The question was... would the economy be able to withstand such an eventuality?

It became necessary in the subsequent years for Indeco to make certain modifications to its takeover programmes. Earlier calculations had discounted elements of business goodwill, payments out of future profits, etc. New elements redefined future payments in assets to be in Kwacha instead of foreign convertible currencies. Similarly, payment for management fees was linked to profits made by the company and to a definite commitment to training of Zambians by the company.

A few years after, the geographical coverage of Indeco was quite extensive and had reached most rural areas. Most notable was Indeco's penetration into rural areas. Commercial activities by area and ownership included a 51 per cent in two oil marketing companies; between 51 and 100 per cent in seven manufacturing companies; 51 per cent in four consumer trading companies; between 51 and 100 per cent in six rural enterprise companies; between 80 and 100 per cent in four property companies; up to 50 per cent in several associated private companies; and 100 per cent in a travel agency company (Eagle Travel). Thus the period from 1969 to 1973, the corporation represented a fairly sizeable stake in the Zambian economy.

Operating through Indeco, the government was able to realise some of its objectives. For instance, it became possible to establish business enterprises which had been difficult to develop. For example, Kapiri Glass Product (KGP) for the manufacture of opaque beer and milk bottles from local material for the domestic market and export to neighbouring Zaire. It created further opportunities for increased indigenous participation in the country's economy. Indeco's shareholding in formerly privately-controlled companies gave impetus to a certain degree of localisation of personnel. This was what facilitated the appointment of many indigenous general managers and/or directors on company boards where Indeco was a business partner. Indigenous appointee performance, within the context of the Mulungushi programme, is a matter for the latter part of this book.

Commercial and industrial activities within the context of the Mulungushi economic reform programme in the rural areas were undertaken by Indeco's subsidiary Rucom Holdings or the Rural Enterprises Group. Its activities were basically joint ventures in partnership with foreign investors. The most favoured were such enterprises as maize milling, timber milling, fishing and country hotels. Selection and location of such enterprises was influenced by prevailing conditions in the particular area. In general, the sufficiency of the market for the service or product in question was justification for pursuing the venture.

Unfortunately, Indeco's rural activities did not break from normal practice. Most of the industries set up in the rural areas were concentrated in the provincial capitals. Many of them also had very weak economic linkages at the production level in their own localities and were not capable of generating much needed employment. For instance, all the bakeries were dependent on wheat flour not either grown or milled locally. Futhermore, management failed to interest itself in rural agriculture projects, such as wheat and rice growing, or establishing of ranches for breeding cattle and other domestic animals which could have supplied the needs of the above industries. The fact that the government had established the Rural Development Corporation to facilitate individual indigenous commercial farmers should not have prevented Rucom from initiating on a commercial basis its own agriculture production units, specifically those intended to support its own production units in the rural areas.

A major policy ommission in Indeco's operations up to 1974 was the lack of country-wide training opportunities for indigenous Zambian covered by its operations.

Mulungushi and Small-scale Financing

The Mulungushi and the Matero economic reform pronouncements addressed themselves to the problem of making credit available to indigenous small scale Zambian businesses. The use of finance in the rural areas became the yardstick for measuring smallness. Lending to and borrowing by indigenous business draws our analysis to aspects or questions of such ambiguous finance and money terms as "liquidity". The concept liquidity may be expressed as the extent to which assets are an expression of money-value. It can also imply the extent to which not only money but can include the maturity of claims to money.

Monetary theory would have us believe that under laissez-faire, the effect on bank liquidity of lending by commercial banks to the government largely depend on the preparedness of the monetary authorities to buy back treasury bills or such other assets as are previously acquired from the government. Similarly, central banks, in this case, the Bank of Zambia, should be able to create money by net credit through acquiring net foreign assets in order to increase banking liquidity and loans to further money-creation. However, even though the above credit creating situations existed, credit was not readily available to small scale indigenous businessmen.

There were thorny problems in the field of credit that could not be solved by Mulungushi and subsequently the Matero economic reform pronouncements. There needed to be changes in the existing structure. The country's financial instruments, markets and the intermediaries for channelling savings and lending were too rigid to accommodate indigenous entrepreneurs. Most banks were foreign-owned and no formal approach had been made to them to find a solution to the problem. The government could not rely on these institutions to ensure the implementation of the Mulungushi nor Matero programme in the absence of some formal guarantees.

Many of the people that the Mulungushi programme wanted to bring into the mainstream were the sort of persons the banks had long avoided as financial risks. It was not uncommon for applications submitted by the former to be turned down. Most of them could hardly prepare a simple balance sheet for the operation they intended to carry out. The very few who got credit were subjected to conditions which many found difficult to fulfill. Expatriate commercial banks always seemed ready to entertain credit-seeking applications from indigenous Zambians and delivering very little in return.[8]

This problem soon caught the attention of the country s policymakers. A few small scale finance institutions, controlled and funded directly by the State, had to be established. Thus, emerged a loan scheme of a comparatively small nature in terms of both scale and allowable individual amounts to be borrowed. It was established alongside other more sophisticated country-wide schemes. The scheme was specifically intended to assist those indigenous Zambians who needed government help not only to start but to stay afloat in business.

The Small Loans Scheme, as it came to be called, was administrated by Indeco on the recommendations from a Committee chaired by the Minister of State in the Ministry of Commerce, Industry and Foreign Trade. Most of the loans approved went toward purchasing established small businesses, rather than towards the establishment of new real investment. The scheme was beset by the constraint of having too many applicants and a very small fund. The lucky applicants were obliged to spend the money in expatriate-owned firms because there were no alternative indigenous firms. This was how after the Mulungushi economic reform proposals, the State established the Zambia National Wholesale Corporation (ZNWC).

Indigenous loan recipients were obliged to make the bulk of their purchases from publicly-owned corporations such as the ZNWC. However, almost everything sold by ZNWC had been imported. Had the ZNWC been a distribution centre for locally manufactured items, the Mulungushi primary objectives of removing foreign domination and increasing local self-sufficiency in production using a greater part of local inputs, could have come to fruition. Commodities which could not be provided by Stateowned enterprises, such as tractors, hammer mills or grinding mills, were supplied by arrangement through E.W. Tarrys, a company in partnership with the Party. The ZNWC innovation therefore failed to break the vicious cycle which permitted the dominance of expatriate firms in the economy.

There were times when the government-sponsored finance houses had to take what in financial circles could only be described as calculated risks. But such action has to be weighed against political expediency at the material time. One example was the loan extended to Timothy Chinunda to complete the Kasama Kwacha Relax Hotel. The action had several demonstrative effects. First and foremost, it was made in the spirit of the MERP. Secondly, as a project, it was likely to generate several economic activities in a very important provincial capital, in part by providing additional accommodation. Hitherto, hotel business had been an exclusive area for European proprietors, apart from the rural hotel at Mansa owned by Luka Mumba. Efforts at securing finance from private finance houses to complete the construction of the hotel had come to no avail and the proprietor would have ended up with an incomplete edifice. The load led to the construction and eventual success of the hotel.

However, the same could not be said for other risks taken by the State finance institutions. The history of the Credit Organization of Zambia (COZ) and subsequently the Industrial Finance Company—a subsidiary of the Finance Development Company (Findeco) contains examples of how to misuse a nation's resources in a relatively short space of time, is a case in point. The Credit Organization of Zambia was established in accordance with its Act of 1967. On 11 August 1967, it took over the assets of the Land Bank which had also proved less capable of sustaining itself, let alone contributing to Zambia's economic development. Both turned out to be white elephants and flourished only briefly as a result of injection of considerable funds by government. COZ was bedevilled by unparallelled inefficiency and mismanagement of funds, especially its inability to monitor the projects to which the loans had been made in the rural areas. Added to the above was the reluctance of the indigenous borrowers to repay. This explains why in the period 1964-65 and the period ending 30 June 1966 the Zambian Treasury contributed £162,766 and £140,824 respectively towards the cost of regeneration of COZ.

Mulungushi and Large-scale Financing

The difficulty of providing money when and where required within the broad principles of the Mulungushi programme was perhaps its biggest problem. Investment finance was required both on medium and long-term bases, especially in the key sectors of the economy. Large-scale financing for economic development was the reason for creating the State Finance and Development Corporation (Findeco), the finance conglomerate. Findeco was given overall responsibility for supervising and coordinating existing and future corporate finance on behalf of the State. However, Findeco, unlike Indeco, did not follow logical guidelines. This was especially shown by its opting to perform certain functions which had caused the COZ to fall.

Findeco's new Chief Executive, Emanuel G. Kasonde, tried to steer the institution in the right course. He displayed not only a great sense of purpose but also attempted to improve the institution's image. Findeco's structure benefited from his long varied experience at the Ministry of Finance. The head office was made as small as possible. Kasonde gave effective supervision by acquiring chairmanship and membership on several boards of the Corporation's subsidiaries.[9] As a holding company, Findeco was comprised of the following subsidiaries:

(a) Zambia State Insurance Corporation Limited (ZSIC), operational up to 1968, with the following subsidiaries: Zambia National Insurance Brokers Limited and City Radio, in which Findeco owned 70 per cent and 50 per cent of equity shares respectively;
(b) Zambia National Building Society (ZNBS), a corporate body of individual depositors but administrated by the Government through its sole ownership of the 800 A shares;
(c) National Commercial Bank Limited, established in 1969. Ninety-eight per cent of the shares were owned by the State. Its task was undertaking normal commercial banking transactions in addition to emphasising its role of assisting in the promotion of agriculture and the development of agro-industries; and
(d) The Industrial Finance Company Limited, a wholly state-owned enterprise, incorporated in 1969, used to rationalize some of Indeco's former lending activities.

Some of the economic reforms incorporated in the Mulungushi and later in the Matero pronouncements were necessary events in the country's search for ways to undo the mistakes of the past and to enhance the welfare of the people. However, the creation of such institutions did not automatically lead to increased financial resources. Apart from a lack of resources, the institutions were also unable to carry out all their good intentions.

MATERO

The decisions taken at Mulungushi or the so called Mulungushi Economic Reform Programme set in motion government's efforts of increasing state control over selected sectors of the economy. The Mulungushi reforms had not significantly contributed to resharping the Zambian economy. Similarly, talk of nationalizing foreign owned banks like Standard, Barclays, and National and Grindlays had faded away. However, in 1969, the Government, through the Matero Economic Reform Programme, took steps to acquire a major controlling equity share in the country's mining sector.

Matero and the Goose with the Copper Egg

Zambia's enormous mineral wealth is best described as a mixed blessing. Mineral income indeed helped to finance some of the country's economic development programmes. On the other hand, this very abundance of minerals was also the main cause of the country's other economic and political problems. This wealth explains the bitter fight waged by settlers during the colonial era to create the Federation of Rhodesia and Nyasaland (FRN). Consequences of the UDI and Zambia's difficulty and costs in devising economic diversity are also directly linked to its tremendous mineral wealth.

The importance of mining to the Zambian economy needs no further elaboration. It is why both the Mulungushi and Matero pronouncements tended to apply a different yardstick to the Mining Development Corporation (MINDECO) than to those applied to the other 24 companies placed under Indeco. In other words, the Government was quite conscious of the need to preserve the Goose which laid the Copper Egg. In trying to assume control over such a crucial sector, it was important to apply tactics which would not backfire.

The Government's interest in having a controlling share in the mining sector goes back to the termination of the mineral concessions and royalty payments to the BSA Company on the eve of independence. Some of the biggest problems faced by the government included the lack of indigenous personnel to assume control over many facets of the industry. In addition, there was the fear of antagonizing foreign investors to a point where the country could be starved of needed capital goods and technical skills. A great deal of money was needed for participation in the industry or even eventual nationalization. These considerations posed real dilemmas. Funds needed to effect nationalization of foreign industrial concerns tend to have a rather high opportunity cost. For instance, external borrowing to repay expatriate companies increases inflationary pressures in most developing countries unless this is offset by increased exports. The question therefore was to ascertain to what extent such funds could be expended, especially in the light of other development programmes.

Nonetheless, the government had already embarked on its course of increased participation in key sectors of the economy. The Mulungushi and subsequently, the Matero spirit had to be seen through. It is therefore inconceivable that the mining sector would be left out of the Matero programme. Part of the background to government action under the umbrella of the Matero economic reform was to stimulate increased investment in mining. Somehow it was hoped that such developments would augment the stream of revenues needed to sustain government expenditures as well as to finance diversification of the country's manufacturing industry and rural growth.[10]

Directing the Mining Industry

No one underestimated the difficulties mineral nationalization would create. Adverse comments when the action was taken, in the wake of the Matero economic pronouncements, were not confined to outside Zambia. Skepticism was also heard within the country and in the local press.

For instance, one local newspaper, on 29 November 1969, alleged that a group of Zambian civil servants and businessmen had thought that the Government had been outwitted when the takeover agreement was concluded. Doubts were expressed about the negotiations which led to the acquisition of a 51 per cent equity participation by the State at the cash price of K209 million, payable to the two companies, Anglo-America Corporation and

Roan Selection Trust (AAC and RST), over a period of eight and twelve years, respectively against 49 per cent left to the expatriate mining companies.

What were the terms of the government and how it was negotiated were the main grey areas on which the nation needed an explanation. According to the Agreement, payment for a 51 per cent share of the combined book value of the mining assets (RST and AAC), calculated as on 31 December 1969, came to US $ 292.6 million (divided in the ratio of US $ 117.6 million (K 84 million) for RST and US $ 175 million (K 125 million for AAC). Bonds issued in respect of payment for shares in RST mines were to be paid over a period of eight years in half-yearly instalments of US $ 9.52 million (K 6.8 million). Those pertaining to shares of AAC were to be redeemed over a period of 12 years in half-yearly instalments of US $ 10.36 million (K 7.54 million).

Provision had also been made in the Agreement for accelerated payments by the Zambian Government to RST and AAC. Accelerated payments would be in 1971 and 1972 to RST and AAC. Repayments were then to be made in each succeeding year. The basic assumption was that Mindeco Limited would be earning about two-thirds of the cost from dividends and that if the earnings were greater than that hypothetical figure, the surplus would be used to speed up the redemption.

The Agreement, as eventually negotiated, enabled the Government to own, through Mining Development Corporation (Mindeco) Limited, a 51 per cent share in Nchanga Consolidated Copper Mines (NCCM) Limited and the Roan Consolidated Mines (RCM) Limited. Directors to each company's board were categorized into A and B groups. Category A represented the majority shareholder (Government) which made every effort to fill its directorship with indigenous Zambians.[11] Category B was made up of the minority shareholders. The remaining shares in NCCM were held by Zambia Copper Investments (Bermuda), a subsidiary of ZAMANGLO (Bermuda). Minority shareholding in RCM was 12 per cent by Zambia Copper Investments (Bermuda), RST International, 20 per cent, and certain public subscribers, 17 per cent.

Changes to the Agreement were conditional upon the Government being prepared to redeem in full, in United States dollars, the outstanding Zambia Industrial and Mining Corporation (ZIMCO)

Bonds (1978) and Loan Stocks. These provisions seem to have been worked into the contract with the aim of protecting minority shareholders' interests.

On the face of it, the Contract Agreement seemed to have vested a wide range of veto rights in the minority, or "B" directors, especially in determining the two companies' (NCCM and RCM) financial policies. However, there is no evidence to suggest that the minority directors ever used this veto power. But it was worrying to know that they could use it when they wished. On the other hand, the "A" directors could, if they thought fit, reject an item on the agenda likely to be considered by each Board.

RST and AAC had also been awarded a management and consultancy contract, in addition to a sales and marketing contract. This may seem inadmissible to the casual reader tracing the logical steps for a country attempting to retain a major controlling share of an important economic sector. The plausible explanation is that it was necessary and unavoidable at the time because of the lack of indigenous expertise. Critics ought to keep in mind the reality of the situation. The government was in no mood to push brinkmanship to a point where it would end up being the loser. The basic reasoning for an economic reform programme was to increase growth and economic welfare, not otherwise.

The newspaper story of 29 November 1969 cast great doubts on the results of the negotiations. So too did the views expressed by scholars like George Simwinga who stressed the adverse cost for the continuity of minority interest, because to him it seemed as if the nature of the contract represented an absence of meaningful government control.[12] Many others also felt that the agreement seemed to have conferred more than commensurate benefits on the minority shareholders over the period before full redemption of the bonds. Other non-Zambian scholars, like Mark Bostock and Charles Harvey, believed that the terms of acquisition could never have been a total victory for Zambian Government given the circumstances and the vested interests of the other partners.[13] The question that had yet to be answered related to the lack of clear elements of effective control by the government over many important levels of the mining industry. To counter some of these weaknesses, the Government resorted to creating institutional machineries like Mindeco. Unfortunately, Mindeco fell far short of proving the needed solutions in part, because its top structure

The Mulungushi and Matero Reforms

was used merely as a forum for debates on policy.

The nation was entitled to some doubt over a deal that affected their own lifeline. Some of these doubts existed in spite of certain changes initiated at the management level. For instance, the Minister of Mines and Industry assumed the chairmanship of the Board of Directors in both NCCM and RCM.[14]

Those who had expressed doubts could not be persuaded to believe that such innovations had got to the root of the problem. In particular, they held the view that it was folly for the agreement to have incorporated payment for management fees on the basis of a percentage of sales. They contended that sales are not a true reflection of profit margins but rather some book value of turnover. In other words, the expatriate enterprise would continue to walk high and dry even when the industry was making losses.

The Mulungushi and Matero programmes in the mining industry left much to be desired. This was because it was an in-between action. Circumstances described above prevented the Government from outright nationalization of the industry. Thus, what transpired was a form of limited indigenization. In this way, the foreign investors (AAC and RST) found it necessary to direct some of their shareholdings to certain domestic economic ventures. The government had at least acquired some leverage in policy formulation over the industry and could monitor how the ensuing profits were spent not only within the industry itself, but also in the economy as a whole.

Perhaps some of the doubts expressed were justified by the events which followed in the train of such action. Acquisition of equity shares resulted in a drastic fall of mining revenues. On the basis of data compiled by the International Monetary Fund, especially its country Report on Zambia, based on Table IV of the International Bank for Reconstruction and Development dealing with Economic and Social data relating to "External Public Debt Servicing, 1969-1989": to redeem the remaining 49 per cent of equity shares, Zambia borrowed US $ 150 million at 13 per cent rate of interest on the Eurocurrency market. Therefore, not only did the government's share of profits from the mining sector fall but were also reduced because of compensation payment to the former expatriate owners. Consequently, the original argument to gain greater control to augment investment was a loss in tax revenues.

Mineral Marketing

The decision to set up and operate a state-run metal marketing agency must have been made as far back as when control of the industry was first conceived. The setting up of the Metal Marketing Corporation of Zambia (Memaco) was a transformation of an old idea into reality.

Establishing Memaco did not change the old structures of the way Zambia's copper was sold on the world market. In other words, Zambian copper continued to be quoted on the London Metal Exchange (LME). The LME is to all intents and purposes a "commodity exchange" and operates as a free market, though subject to various constraints. The prices quoted not only vary but are determined by processes corresponding to a continuous auction of the commodity in the market-place.[15] Since 31 May 1968, spot quotes are made on the same market. Zambia's main problem is the lack of influence in the world market, especially where most of her primary minerals (copper, cobalt, zinc and vanadium) are sold. This problem is not peculiar to Zambia alone but is general to almost all developing countries. The only sign of influence is exerted indirectly. For example, in times of strikes. Unfortunately, this conferred no material advantage to the Zambian economy. The rise in the price of the metal in light of a Zambian strike benefits other copper producers. It also makes possible the danger of eventual loss of the old markets to substitutes like aluminium products.

There are certain developing countries in a much worse situation than Zambia in respect of how their commodities are sold in international markets. For instance, raw diamonds from an African state are transported for sorting and pricing to a firm in the "Middle East" before being sold on the open market. It is quite likely that the African country is permanently defrauded of the real value of its diamond export revenues. One need not be a genius to see that it would be considerably cheaper to grade the diamonds in the country of origin. However, the lack of facilities to perform that function locally has been one reason for the continuation of that sad state of affairs. With such a change, the declared value would then enable the treasury to compute the appropriate customs duty.

What was hoped for under the MEMACO structure was for the state to monitor the trend of sales and to provide market intelligence for production projects. MEMACO was like most structures in formerly colonized countries. It followed in the footsteps of its predecessors, AMAX, in all of its deals. There is very little evidence of any attempt by MEMACO to reverse traditional trends. MEMACO was established to operate under conditions of "mixed market". A "mixed market" is a mixture of different structures. It includes oligopsony (where a small number of importers prevail), closed market (where exporters and importers are associated on long-term basis through joint ownership or through both horizontal and vertical integration), controlled and protected market (where a major part of commodities is under permanent state control). MEMACO would have been more successive had it operated under conditions of closed market rather than that prevailing under both the mixed and free market.

The objective rather than the experience of MEMACO (Zambia), in the period under review, is atypical of national efforts in most Third World countries in the areas of marketing of primary commodities. This is particularly applicable to marketing in the traditional markets. Certain Third World countries on the other hand, and especially where the production patterns have not enabled them to penetrate traditional markets, have put efforts into finding alternative markets. These countries have concentrated their efforts in marketing into non-traditional markets. Examples of these latest developments, most of which are a post-independence creation include: CODECO, the Copper mining and exporting company of Chile; CVRD the Brazilian State-controlled mineral producing and exporting company; COMUNBANA, the marketing arm of the Union of Banana Exporting Countries (UPEB). A common denominating factor is that each owns and controls sales offices or trading houses or regions in which it is respectively active.

Unfortunately, the positive character of the above initiatives, including that of MEMACO, have to a large extent been derailed by several contradictory international trade factors. These include a lack of a coherent and coordinated global strategy among producers in the Third World countries. For instance, for some copper exporting countries, not even in the creation of the Council of Copper Exporting Countries (CIPEC) have such global strategies evolved with any commensurate success. Over and above, there is a definite failure by many of them to attract international support and finance, necessary for the survival of these institutions. Their preoccupation has thus remained one of survival. And this has compelled most of them to adopt piecemeal approaches. In particular, a lack of specific programmes of financial and necessary technical supporting facilities. In addition, the structures of these institutions are not geared to funding the initiatives.

Perhaps stress at the national level should have been put on increasing technical and financial support with a view to the development, not only of marketing, but also production and the distribution systems taking account of the difficulties they encounter in respect to price fixing in the international markets, a situation which is conditioned to a greater degree by speculation, hedging and the like. Consequently, these institutions will need to review their strategy in terms of improving surveillance.

THE REFORMS AND INDIGENIZATION

The reforms established committees to oversee the implementation of Zambianization. Zambianization is synonymous with indigenization or Africanization. The *raison d'être* of indigenization as pointed out in the previous chapters was elimination of class categories in commerce and industry: Europeans at the top; Asians and other non-African minorities in between; and Africans at the bottom of the ladder with occasional overlapping for the two upper groups. The group categories observed above were also not much different to the civil service structure of most Africans countries at independence. For example, expatriates dominated the administrative and professional posts with Africans confined to clerical and other menial duties. The only exceptions being Nigeria and Ghana where Africanization had already gained momentum before independence.

The concept of indigenization has been defined by experience to include staffing of the State government and the various sectors of the economy with indigenous Zambians, and their control of the management thereof. A continued preponderance of expatriates in the country's key economic sectors implies peripheral role by the State and its people in directing policy. This is not to deny the fact that expatriates had a vital role to play in the country's process of economic development. However, it stood to reason that the private and public sectors should be run by indigenous personnel, like the cabinet and the government ministries.

The reader will recall earlier mention of termination of the Witwaterstrand Native Labour Association (Wenela) contract through which Zambian native labour was recruited to work in South African mines. Previous Zambianization Committees dealt separately with the mining industry, the private and parastatal sectors. All three were replaced by one committee in 1971. The terms of reference for the new committee included among other things, that it:

> (*i*) undertake or initiate studies and investigations into the progress of Zambianization in any firm, organization or industry in the private and parastatal sectors of the economy;
> (*ii*) review training arrangements designed to facilitate the replacement of expatriates by Zambian personnel, and to make recommendations for the development and improvement of these arrangements;
> (*iii*) review the availability of Zambian manpower for employment in the private and parastatal sectors, particularly in relation to positions held by expatriate personnel, and make recommendations as to how Zambian manpower can be improved to facilitate Zambianization.[16]

The government's regulatory labour laws should have made it incumbent upon the mining companies to initiate appropriate training programmes aimed not only at making indigenous Zambians good miners but also integrating them into the management structure. A specific time-frame should have been set. The same stipulation should have been given to other major expatriate businesses in the country.

There is evidence that the mining companies had been attempting some training programmes. However, those programmes were based on their own immediate business interests. There was a lack of long term planning. One official mining publication stressed some of the achievements of the mining companies in this field. The report indicated the policy towards training and education within the industry: The Company not only maintained their long standing scheme of industrial training, but also increased the emphasis on a number of them, and in many aspects broke new ground in 1966 by launching additional training and educational facilities. Further to the 1965 programme of training for promotion and Zambianization these additional schemes meant that prospects for local employees were appreciably enhanced. The number of employees given formal education or training during the year was in the region of 15,000 at a cost to the industry of some three to three-and-a-half million pounds.[17]

Statements of this kind convey a sense of self-congratulation, in the absence of other facts. In reality, one could observe several gaps as regards to the calibre of the trainees and the trainers. Like almost all self-praise statements, especially those intended to deceive, it emphasised important issues but not matched by tangible results. Emphasised in respect of the above were:

(*i*) training for promotion;
(*ii*) learnership in occupations as wide ranging as surveying, assaying, accounting, etc;
(*iii*) selection and placement, which was being undertaken with the assistance of the Government's Educational and Occupational Assessment Unit;
(*iv*) adult education, including scholarships and other forms of assistance given to enable selected individuals to attend courses of vocational training at colleges and universities both in Zambia and abroad; and
(*v*) induction training of new workers.

The extent to which this training had taken effect can be judged against later results. Previously, it was easy to blame this absence on the influence exerted by the South. But this should not have prevented the companies from taking the initiatives given the special Zambian conditions. A start could have been made a few years after independence and before the Matero action with the appointment of at least one administrative assistant at one of the many mineheads or refineries. However, only the RST found it possible to appoint only one indigenous Zambian director, Alderman Safeli H. Chileshe, on two out of six of its boards of directors.

It is useful to compare the situation in 1966 to that of 1973 upon takeover.[18] The mining operations of both AAC and RST were by 1966 quite elaborate. Each system of management consisted of a Chairman, Board of Directors and respective administrations of the mines at the minehead and at the refinery level. Six separate administrative groups had been set up with a lot of interlinkages created at the director level. Some directors of one group held directorships on the boards of their counterparts. It was the rule rather than the exception that there was a total absence of indigenous Zambians at the senior levels of administration. Influence exerted by South Africa was probably one reason why AAC did not take a similar action. This was puzzling in view of the strong leadership and courage of the South African Sir Harry Oppenheimer, regarding industrial structure changes. Sir Harry's forthright attitude for change was made abundantly clear in his statement delivered in London in May 1976.[19] Had indigenous Zambians been appointed or served on the various boards of the expatriate companies and if indigenous Zambians had been promoted into learnership of surveying, assaying etc., implementation of the Matero programme would have been relatively easier.

The predominance of indigenous Zambians as directors on the various boards of the two mining companies by 1973 had made a major inroad. It occurred because of increased government participation in the mining industry through the acquisition of 51 per cent equity shares. A most notable development in the mining industry by 1973, apart from the increased spate of indigenous directors, was the ease with which the post of company secretary went to indigenous Zambians. The fact remains that these actions should have been taken on a voluntary basis, from the start, by the expatriate mining interests themselves.

The extent and speed at which indigenization goals would have been met by the expatriate companies without nationalization is difficult to establish in precise terms. They were only marginally effective in terms of localization of their administrative posts. Their training and investment programme of 1966 should have shown some tangible results in Zambianization, following the MERP and Matero programmes of 1968 and 1969. The appointment by RST of its first indigenous assistant managers of administration at some of their mines should have spread to other areas. AAC's sending three indigenous Zambians abroad on a special training programme should have become a regular practice. Unfortunately, none of these things occurred. It is regrettable that the Corporation was unable to even retain the personnel trained by itself, all three had to find employment in other quite unrelated sectors of the Zambian economy.

By and large, we have concentrated the study of indigenization on the mining sector. The policy of indigenization of skilled manpower in other sectors sometimes caused replacing expatriates with less skilled indigenous labour. Where this took place, the end result was a decline in efficiency, which resulted in a drop in short-term national income and a fall in the long-run rate of economic growth and development. This effect is explored further in chapter seven.

THE REFORMS REVISITED

The way to analyse the MERP and Matero programmes is to review the extent to which the objectives were achieved. The preceding account of developments in key economic sectors shows successes and failures. The experience provides numerous lessons for the succeeding decades. For instance, direct government intervention can either complement or thwart domestic efforts, as in the cases of the Credit Organization of Zambia, Namboard and Zamcab, etc.* Hence, the effectiveness of reforms is contingent upon the structure of the economy itself and the nation's political philosophy.

Direct investment in mining by private minority shareholders (AAC and RST) was required as long as there were prospects for returns. The development of agriculture through direct state involvement to meet both domestic demand and to expand the foreign exchange base proved rather difficult. Zambia's abundant agriculture potential came no closer to full realization under the reform programme. In point of fact, the rural-urban terms of trade tended to shift against agriculture, notwithstanding increased official producer prices.

There were several constraints that affected rapid industrialization, which was the policy adopted under Mulungushi. It was also a policy to emphasise import substitution. Policy-makers did not consider the fact that such industries were largely dependent on imported inputs. The list of constraints included the lack of

---

* Zambia's northern neighbour. Tanzania discovered later, and much greater cost to itself that it had made a big mistake when in 1967 it nationalized the country's sisal industry.

industrial institutional infrastructure, difficulties of mobilizing financial resources and the lack of industrial and technological information. It is a prerequisite in respect of industrial development for planners and operators to have readily available a supply of information on which to develop or from which to base the choice and acquisition of industrial technology. However, Zambia seemed to have neglected this part of their programme of industrial programme.

Almost the same sort of constraints were encountered in respect of the development of the country's agriculture sector. Many ideas had been put forward aimed at solving some major agricultural problems like increasing production and improving distribution of products. For example, direct government action was initiated and supervised through various ministries, coordinated by the Ministry of Agriculture and Rural Development (later renamed the Ministry of Agriculture and Water Development). Indirect government action was taken through various parastatals like the National Agricultural Marketing Board (NAMBOARD). The Dairy Produce Board (DPB), the Agriculture Rural Marketing Board (ARMB), the Rural Development Corporation (RDC) etc.

However, increasing of prices proved ineffective in solving these problems. This was the case in Zambia as also in other countries, like Kenya and Mali as the European Economic Community realized with many of their aid programmes. There is no denying that higher prices to farmers can be a major incentive for increased food production. But the Zambian experience showed that it was by no means a sufficient condition. Price increases need to be supported by other complementary factors. First, as stated by J. Verloren van Thermaat, there had to be an optimal price mix between the various products. Rather than the absolute price of a crop or produce, say maize, tobacco or milk, other things must be used as an incentive to produce. It has to be the relative price of that crop compared to other crops like cotton, groundnuts etc. Second, there had to be an improved availability of inputs such as fertilizer, insecticides, water and technical know-how. Third, there had to be a major improvement in appropriate applied research on a decentralized level. That research must be aimed at significantly improving yields both per hectare and per manhours of labour.[20]

We have demonstrated, in the case of Korea and Japan, that

throughout recent history, research has played a crucial role in increasing yields. Consequently, government intervention either by indigenization or through ministry or parastatal action was required where the market forces of supply and demand could not alone solve the problems. Concomitant with the catalytic role of research are facilities for disseminating the research results.

Neither complete success nor failure was fully realized in the sectors, which formed the basis of our analysis of the reforms. Theory stresses that every wrong has its remedy and that an appropriate solution is not beyond the ingenuity of man. However, long-run prescriptions to correct wrongs were dismissed with cruel brutality by John Maynard Keynes, who said that most people do not live to see long-run solutions. Viewed in this broader context, the shape of things which followed in the wake of Mulungushi and Matero reforms were short-term and only partly a result of government action to enforce them. The biggest challenge of the period was how to bring about major long-term changes in all sectors of the community. The size of the problem was perhaps the dilemma which afflicted Mulungushi and Matero economic reform frameworks and consequently, frustrated the Government.

The preceding analysis is far from being exhaustive, but it justifies the conclusions. It also attempts to flag some of the problems and attempts at solutions in the field of economic development.

REFERENCES

1. K.D. Kaunda, *Towards Complete Independence*, Zambia Information Services, Lusaka, 11 August 1969, p. 4.
2. See the Harrod-Domar model especially Evsey Domar, "Expansion and Employment". *American Economic Review*. March 1967, pp. 34-35: As a theory it relates a country's rate of growth of income to its savings-income ratio and marginal capital-output ratio. However, this theory was evolved against the background of an advance economy unlike that obtaining in the Third World Countries. This is because the theory tries to explain how much income has to grow in order to induce sufficient investment to maintain a certain desired rate of growth in income.
3. Jonathan H. Chileshe, "Aid: The Amorphous African Development Catalyst", *Symposium on African Perspective on the New International Economic Order*, United Nations University with Addis Ababa University, Addis Ababa, 3-9 May 1981.

4. Alderman Safeli H. Chileshe, served on the Boards of RST, Barclays Bank, Tanganyika Investment and BAT; similarly, Alderman Tom Mtine served as a Director for Lonhro and BAT.
5. *Annual Report of the Department of Labour for the Year 1976*, Ministry of Labour, Government Printer, Lusaka, 1968, p. 11.
6. E.A. Kashita, "A Policy of Rapid Sustained Industrialization" *Enterprise*, No. 4, Zimco, Lusaka, 1974, p. 7.
7. K.D. Kaunda, *Zambia's Economic Revolution*, Zambia Information Services, Lusaka, 19 April 1968, p. 46.
8. Jonathan H. Chileshe, "Indigenization", *Enterprise*, No. 2, Zimco, Lusaka, 1975, p. 43.
9. Falcon Group, *Seven Years of Progress*, Review Promotion Ltd., Ndola, Zambia, 1972, p. 73.
10. K.D. Kaunda, *Towards Complete Independence*, Zambia Information Services (Matero) Lusaka, 11 August 1969.
11. Ph.H.A. Brownrigg (formerly of Anglo-American Corporation) was the only exception. He was nominated as a director on the Board of NCCM along with L.M. Lishomwa, F.C. Sumbwe, L.J. Mwananshiku and W.M. Chakulya.
12. George K. Simwinga, "The Copper Mining Industry of Zambia", *What Government Does*, edited by Mathew Holden Jr. and Dennis L. Dresang, Sage, London, 1975, p. 87.
13. Mark Bostock and Charles Harvey (eds.) *Economic Independence and Zambian Copper: A Case of Foreign Investment*, Praeger Publishers Inc., New York, 1972, pp. 169-173.
14. Hon. H. Mulemba, MP., as Minister of Mines and Mining Development was the first indigenous Zambian to become chairman, although the negotiations resulting from the announcement of 30 April 1973, regarding Government participation, were not yet completed at the time. Hon. Elias A. Kashita, MP., as Minister of Mines and Industry, succeeded Mulemba as chairman.
15. Detailed description can be found in Blau, Gerda, "Some Aspects of the Theory of Futures Trading", *Review of Economic Studies*, Vol. 12. 1944; C.W.J. Granger (ed.), *Trading in Commodities: An Investors' Chronicle Guide*, Goodhead—Faulkener, Cambridge, 1975.
16. *Annual Report of the Department of Labour for the Year 1971*, Ministry of Labour and Social Services, Lusaka, 1971, pp. 4-5.
17. *Zambia Mining Industry Year-Book 1966*, Copper Industry Service Bureau Ltd., Kitwe, pp. 24-27.
18. *Zambia Mining Year-Book 1973*, Copper Industry Bureau, Kitwe.
19. Harry Oppenheimer, "The Fifth Stock Exchange Chairman's Lecture" delivered to the Stock Exchange, *Supplement to Optima*, one.
20. J. Verloren van Thermaat, "No Trade, Less Aid: The EEC's New View on Its Relations with ACP Countries", *Lome Briefing*, NGOs to the European Communities, Brussels, No. 13, Column 6, 1983.

Chapter Six

Parastatals

The administrative structure of numerous developing countries is, and remains to this day, one that was borrowed from their colonizers during certain periods of their histories.[1]

A parastatal is taken to mean any public corporation or a state trading organization in which the State owns a considerable measure of its operations. In other words, these are commercial enterprises to which governments delegate some of their authorities. Experience in most developing countries is relatively recent but they follow a similar administrative pattern. Their evolution comes from the concept of national economic transformation, a product of the industrial revolution which was used extensively in the developed economies.

A definition of a public corporation is an institution operating a service of an economic or social character, on behalf of a government, with a largely autonomous management, even though the corporation is responsible to the public through government and Parliament, and subject to some direction by the government. It is equipped with independent and separate funds of its own and with the legal and commercial attributes of a commercial enterprise.[2]

No matter what definition of a parastatal is adopted, their objectives remain the same. They are government instruments for intervention in the various sectors of the economy. Public corporations are distinguishable from private enterprises because they are a creation of governments. The legal features of state trading organizations distinguish them from traditional organizations such as a Ministry of Trade or Industry.

It is important to avoid defining state enterprises by stressing the demarcating line between them and the state, as if they were two separate arms of the same state machinery. Rather, attention ought to focus on how each can be seen to contribute to overall economic development without too much overlapping of their respective functions.

Perhaps the main cause of confusion over the definition was inevitable given the Zambian economic setting after independence. Such confusion was encountered in the implementation of activities pertaining to integrating commerce, industry, finance and agriculture.

The emergence of and role played by such enterprises in the economies of developing countries varies only slightly from country to country. In Zambia, as in neighbouring Tanzania or Kenya, these institutions have a definite legal status and are supposed to eventually control most of the country's foreign trade on behalf of the State.

Public corporations are a direct result of growing nationalism in many developing countries. The emergence of public corporations was perhaps the most important occurrence during the twentieth century.[3]

Some of the vexing issues in regard to the emergence of these state enterprises included the notion of economies of scale which it was thought should be basic to the activities of state or public enterprises. In the period under review the Zambian market was too small to support the large planned units unless a way was found to link them together. Similarly, the purchasing power of the nation also was a limiting factor and could not be relied upon to support the large-sized enterprises capable of exploiting the inherent advantages of economies of scale.

THE CONCEPTUAL FRAMEWORK OF PUBLIC CORPORATIONS

The *raison d'etre* of indigenization was as stated in the previous chapters the desire by the new governments to remove and eliminate pre-independence economic inequalities. With independence, African governments relentlessly pursued a policy of Africanization (indigenization) of their respective civil services. These actions were not weighed against the cost of loss of efficiency and performance effectiveness. Rather, by consideration of the only viable political alternative. It was only to be expected that the African governments would soon deem it feasible to turn their attention to Africanization or indigenization of the key sectors of their national economies.

Another reason for parastatals in the post-independence era was the inherited colonial economic structures. Consequently, the creation of public corporations represented a defence against the dangers of foreign economic domination. Through these institutions the state tried to acquire capital and the necessary means of developing alternative, indigenously-owned and managed businesses in the economy. They were also to be watchdogs over developments and economic activities which prevented any small local clique from cornering for itself certain benefits to the exclusion of the nation as a whole.

The conceptual framework of public corporations or state trading enterprises stems from the economic development objectives expected of them. These institutions are viewed in many countries of the Third World as a channel which the state can use to intervene in the process of development. It is not so much that developing countries are somehow obsessed with the idea of catching up with the developed countries.[4] On the contrary, the framework and justifications for the emergence of public corporations can be summarised as follows:

(*i*) promoting self-reliance in strategic sectors of the economy;
(*ii*) providing the necessary infrastructure facilities for a country's balanced and diversified economic structure;
(*iii*) reducing regional disparities in development;
(*iv*) allocating and managing a country's resources efficiently and prudently channelling foreign capital and local funds into productive sectors;
(*v*) preventing concentration of economic power in the hands of a few individuals;
(*vi*) contributing to employment generation and labour productivity in all sectors of the economy; and
(*vii*) reinforcing social control on trade and industry in order to ensure equitable distribution of goods and services between the urban and rural areas.

This background leads to state enterprises being vested with both the social and economic responsibilities for ensuring progressive and equitable distribution of any surplus generated from their productive activities.

Developing countries have shown great impatience over

acquiring technology, capital and managerial skills from abroad but don't want to compromise their cherished freedom and independence in their pursuit.[5] Public corporations were considered one of the ways of retaining such independence. It is often assumed that through public corporations a government could increase and improve its influence and bargaining power, especially in dealing with transnational corporations.

The emergence of these institutions is to a large extent, a product of political decisions against an inherited economic structure. It is also a fact that the state, not only within Africa but the world over, has emerged as a major engine of both growth and development. In other words, only concentrating on such traditional matters as the maintenance of law and order, collection of taxes and providing social and physical facilities was not enough. Something had to be done directly and indirectly by the government in order to get things moving in the right direction.

During a Reith Lecture on the British Broadcasting System in November, 1966, J.K. Galbraith stated that state intervention was a common phenomenon even in avowedly capitalist countries like the United States of America. The notion of free enterprise has become a minor branch of theology. In other words, the state is playing an important and increasingly larger role in economic affairs. This is true even in the supposedly unplanned and capitalist economies of Japan, Taiwan and the Republic of Korea. Later in this analysis, an attempt will be made to show that it is commonly accepted that a government of a given country can directly control certain operations and/or reduce or eliminate economic activities which fail to provide linkages to other sectors of the economy.

LEGAL AND SECTORAL ANALYSIS OF ZAMBIA'S PARASTATALS

An understanding of certain Zambian laws can elucidate the country's policy towards the establishment of public corporations and their relation to the process of economic development. Legal structures provide a background to the issue and knowledge of them helps sharpen appreciation of the origin and role of parastatals. This analysis will show the extent to which the government was able to use laws and legislative machinery in support of its actions and policies. Reference shall be made to the following laws:

(i) Mines and Minerals Act, Chapter 329;
(ii) Mines Acquisition Act 1973 (Special Provision Amendment);
(iii) National Agriculture Marketing Act, Chapter 356;
(iv) The Companies Act, Chapter 686;
(v) The Trades Licensing Act, Chapter 707;
(vi) Pioneer Industries (Relief from Income Tax) Act, Chapter 666;
(vii) Industrial Development Act No. 18 of 1977; and
(viii) Statutory Instrument Number 41 of 1969.

The analysis will emphasise policy sections of singular relevance to economic reform, in which the public corporations were to be the machinery for achieving set objectives. The analysis will investigate whether informed legal interpretation was justified in vesting state trading organizations with so much power.

The Mining Sector

The dominance of mining in the Zambian economy is well known. Mining as a sector was first in importance even to sectors of services and infrastructure in terms of sectoral distribution (percentage) of gross domestic fixed capital formation during both the FNDP and the SNDP. Between the periods of 1966-1970 (FNDP) and 1972-1976 (SNDP) it accounted for 26.4 per cent and 29.0 per cent of Gross Fixed Capital formation. Other Sectors of the Zambian economy provided much less of the Gross Capital formation.

Zambia by the Matero announcement of 11 August 1969 acquired 51 per cent interest in the mining industry. The first question to ask is why did the new government take so long before using its legal muscle on the mining sector? As a matter of fact, the representatives of the mining giants like Sir Ronald Prain of RST had indicated in 1965, and also in 1967, that a great possibility existed for governments in developing countries taking command of their sensitive economic sectors.[6] In 1965, during the Eighth Commonwealth Mining and Metallurgical Congress in Melbourne he cautioned that developing countries, at one time or another, would inevitably have to exert direct influence in the copper industry.

The emergency and timing of public corporations covering the Zambian mining sector ought to be viewed beyond the pivotal role of the industry in the economy. The timing of government action was constrained by lack of domestic resources, both human and otherwise. There was the dilemma of resource development in respect of nationalization, indigenization options in the face of other competing economic development needs. However, the most compelling reason for finally taking the move was Zambian disenchantment with the degree of investment in the sector by the principal mining companies.

The government observed that there had been very little mine development by the private mining companies apart from that at Kalegwa, Mimbula Fitula and the expansion at some existing mines. The government could no longer accept the argument being advanced by the mining companies, especially their claim that the royalty system was an obstacle to their future investment policy.[7] This polarization of views did not immediately lead to the creation of a public corporation in this sector. Negotiations between the Zambian Government on one hand and the Anglo-American Corporation Group and the Roan Selection Trust Limited, on the other commenced immediately following the Presidential announcement.

Effective Ist January 1970, the mining undertakings (assets and liabilities) of Rhokana Corporation Limited, Nchanga Consolidated Mines (1937) Limited (under its then name of Nchanga Consolidated Copper Mines Limited) and Rhokana Copper Refineries Limited were merged with mining undertaking of Bancroft Mines Limited to form the Nchanga Consolidate Copper Mines Limited (NCCM). The NCCM's capital structure was reorganized and through a wholly-owned subsidiary (Mindeco) of the Zambia Industrial and Mining Corporation Limited (Zimco), a State Corporation, acquired a 51 per cent interest in the equity capital, and Zambia Copper Investments Limited (ZCI), a company incorporated in Bermuda, acquired the remaining 49 per cent. ZIMCO issued to ZCI loan stock 1982 as payment for its 51 per cent equity in NCCM.

The Mining Development Corporation (Mindeco) emerged with a wide mandate and was intended to facilitate among other things:

(*a*) expanding local secondary manufacturing industries which could make maximum utilization of primary products derived from the mining sector;

(b) encouraging the manufacture of equipment and tools within Zambia to meet mining requirements;

(c) continuing measures aimed at diversifying into other minerals apart from copper, such as lead and zinc, and setting up small mines;

(d) ensuring that indigenous business prospecting organizations acquired sufficient technology and the capability to compete for contracts locally and internationally;

(e) ensuring that the Geological Survey Department accelerated regional mapping of the whole country in order to provide the necessary data upon which investment decisions could be based;

(f) intensifying the prospecting programme for oil;

(g) ensuring and striving for a realistic pricing mechanism for all of the country's minerals; and

(h) adopting measures and formulating necessary legislation to ensure that:

(i) reasonable incentives existed to speed up the opening up of new mines;

(ii) there was adequate conservation of Zambia's mineral resources and that wasteful methods of mining were prevented;

(iii) Zambia's environment was adequately safeguarded; and

(iv) the training of Zambians and Zambianization in the mining industry and the Geological Survey Department were accelerated.

It was not until almost six years after independence, in January 1970, that the Parliament enacted Chap. 329: the Mines and Minerals Act, to facilitate implementation of the above tasks. The principal aim was stated to be the regulation of all laws relating to mines and minerals and provide for the granting of, renewal and termination of mining rights. The Act also established a Mining Affairs Appeal Tribunal and defined its jurisdiction over matters connected with or incidental to what was set out in its sections and sub-sections.

The newly created structure, Mindeco, was made responsible for ensuring the implementation of the reforms and intensifying extensive exploration and exploitation of the country's mineral potential. Added to the above was the task of controlling the 51 per cent government interest in Roan Consolidated Mines (RCM) and the Nchanga Consolidation Copper Mines (NCCM). Subsequent amendments to the Act in 1973 were intended to strengthen the operational structure of Mindeco, which became a subsidiary of the Zambian Industrial and Mining Corporation (Zimco), even though it remained a holding company for RCM, NCCM, Kafubu Emeralds, Zambia Broken Hill Development, Ndola Lime, Copperbelt Power and Mines Air Services.

In the course of these developments, the Metal Marketing Corporation of Zambia Limited (Memaco) was created as a public corporation to act exclusively as a sales agent for NCCM and RCM. This public enterprise was also incorporated as a wholly-owned subsidiary of Zimco on 25 September 1973, with the mandate to market all metals and minerals produced in Zambia. Memaco Services Limited, a subsidiary of Memaco, was incorporated in the United Kingdom on 5 August 1974, to provide marketing and sales services which the parent company could not readily perform from Zambia. The work envisaged and carried out under the auspices of Mindeco Limited was often buttressed by appropriate legislation and complemented by action at the level of the Ministry of Mines. For instance, the decision to invest K 9 million in coal mining was part and parcel of activities of the Metals and Minerals Development Unit in the Ministry of Mines, undertaken in close cooperation with the development of geological surveys.

The Commercial and Industrial Sector

Zambia's indigenization policy in commerce and industry differed from that observed in Tanzania and was much nearer to the pattern adopted in Kenya. It was less interested in taking over foreign-owned enterprises unless compelled by circumstances. Rather, it was interested in increasing government and African participation in new economic expansion. Integrating the commercial and industrial sectors which were predominantly in the hands of expatriates was a very difficult exercise in Zambia. Development problems in many poor countries are not only central but also involve human aspects. The adverse comments to the announcement of April 1968, about directed government action in this field could not have surprised any one, least of all in Zambia. The announcement was considered repugnant because they were seen as upsetting the "applecart" or the *status quo*. The initial reaction was to be expected because the reform introduced a totally new element in the economy. This upset them because earlier, expatriate commercial communities were the only ones who had determined economic and commercial parameters without such interference by the state.

In earlier analysis, mention was made of the lack of indigenous know-how necessary to cope with the new situation. Consequently, the government turned to an existing structure, i.e., Indeco, in its attempt to mobilize the necessary leverage. Indeco and all its subsidiary companies acquired 51 per cent shares in 24 companies involved in retailing and wholesaling, builders' hardware and tools, ammunition and arms, bakery products, furniture and householding fittings, curios and antiques, meat and fish products, carpets and linoleum, agricultural supplies and fertilizers, jewellery and watches, and coal and coke.

It was fortuitous that the takeovers based on the Mulungushi Programme proceeded fairly smoothly both from the point of view of private owners and Indeco itself, which appeared satisfied with the results. This smoothness was in part attributable to new legislation and the flexible line adopted by the government in implementing the actions. There were inconveniences suffered by consumers in some rural areas. Non-Zambian traders (mostly Asians) were forced to close their businesses but then Indeco and its subsidiaries failed to provide immediate alternatives for the consumers who had hitherto depended on the Asian merchants.

The promulgation of legislation, in particular Chap. 707 of the Trades Licensing Act and subsequently Act No. 18 of the Industrial Development Act of 1977 allowed Indeco and indigenous African businesses to take over from expatriates. The amended Trades Licensing Act provided for matters incidental to or connected with trade or commerce. This was followed up by the Ministry of Trade and Industry issuing, in January 1972, a number of guidelines intended to inform the business community of the body and spirit of the economic reform programme itself. The government blueprint was another attempt to reassure those being replaced of the government's wish to forge some form of partnership in bringing about overall economic development.[8]

Parastatals

Statutory Instrument No. 41 of 1969 amended Section 17 of the Trades Licensing Act in line with the decisions taken both at Mulungushi and Matero by restricting the issuance of certain licences to non-citizens. However, an amendment to Sub-section 1 of Section 17 of the same Act did permit licensing authorities to issue certain restricted licences to non-Zambian citizens with the consent, in writing of the appropriate Minister. Provisions under Sub-section 2 of Section 17 of the Act stipulated some of the criteria in dealing with reserved licences which could be awarded to a corporate body or to a partnership, of which a partner was a corporate body.

Promoting industries in Zambia was difficult in the face of former links with Southern Rhodesia, especially the fact that the latter had been taken for granted as Zambia's supply base. Fostering and developing industries through Indeco had to be enhanced with the help of appropriate legislation. The task of implementing the programme also required effective financial support not only to the various subsidiaries of Indeco but also to those few enterprising Zambians able to carry out industrial activities. It is important to note that the legislation attempted to provide for licensing and control of various economic activities and provision was therefore made to grant incentives and to attract investment. The architects of the same legislation also provided for procedures by which contracts could be drawn up with Indeco for the transfer of foreign technology and expertise.

The tendency of buying up expatriate commercial interests did not necessarily confer effective indigenization. It only represented a change in ownership of the same stocks. Post-independence Kenya had rejected such a course of action because it was argued, and rightly so, that money paid for compensation to aliens for nationalized assets would inevitably leave the country. Consequently, some of Zambia's increased foreign exchange problems were in part contributed to by some of these developments.

In retrospect, certain important elements were missing, which would have made implementation of this process easier. The major missing link was not providing for human skills, which were no less important than the availability of investment funds and industrial raw materials. Greater efforts should have been made to attain the goals of manpower self-sufficiency in strategic sectors within a pre-determined time-frame. The perpetuation of the lack of skills would significantly affect not only the policy of indigenization but also the margin of savings and investment in the country, especially that derived from domestic resource mobilization and job opportunity creating.

The Money and Finance Sector

The history of commercial banking in British Central, Eastern and Southern Africa started on similar lines to what went on in British West Africa. It was basically associated with the establishment of branches of Barclays and Standard banks. These two banks had close links with their head offices in the United Kingdom. However, their entry into Zambia was through South Africa. Their policies were determined by their parent companies and could not necessarily be adjusted to the new situations.

The move to create appropriate parastatals in the field of finance and banking and the ensuing legislation, to a large extent, followed similar lines to those taken with regard to other economic sectors. Most important of the institutions to emerge in this field were the state-owned National Commercial Bank, the Development Bank of Zambia (DBZ), the State Finance and Development Corporation (Findeco) and insurance businesses. The original objective for creating the National Commercial Bank of Zambia had been to break the hold of existing expatriate banks. However, when this line proved unattainable, the state adopted a policy of mutual coexistence. It is not so much the history of their creation which is important but the purpose for which these institutions were expected to serve. Mobilization of domestic and foreign financial resources for the implementation of the reform programme were of paramount importance and was the deciding factor between success or failure of the programme.

The old institutions acted as financial intermediaries for channelling funds from lenders to borrowers. But they could not be expected to respond with the same speed as those created by the government itself. It would therefore have been quite difficult to bring about rapid changes in the absence of certain drastic changes in the money and financial field. As it turns out, it was the continued existence of the expatriate banks which helped the process, by raising part of the country's bridging finances, mobilized both domestically and from foreign markets. The concept of appropriate financial institutions derives in part from what we have just described. Basically, expatriate banks remained in order to provide money to meet required capital investment.

Parallel with the support received by the government from the expatriate banks was the creation of the National Commercial Bank Limited. The government owned a 60 per cent controlling interest in the bank when it became operational in 1970. It was expected to extend its branches into the semi-urban and rural areas in order to spearhead the relaxation of traditional lending criteria, especially to indigenous businessmen. Hitherto, banking facilities had been confined only to the urban centres. The rural sectors were served by post offices, not as lending institutions, but as places to receive small savings as deposits.

Alongside these developments were actions initiated by the Bank of Zambia, which accepted the role of being the lender of last resort. The legal structure of the Central Bank provided for control over credit and exchange. It could thus facilitate the redirection of financial resources through the combined efforts of the Central Bank and the commercial banks. It could ensure that business was conducted in a financially sound manner and that profits generated were ploughed back in the economy, and not merely distributed to shareholders for externalization.[9]

The creation of Findeco and the various structural changes introduced thereafter could have reinforced the implementation of the policies enunciated in this field. Findeco's structure as a parastatal under the Ministry of Finance in April 1971, was that of holding company for: J.H. Minet (Zambia) Limited, an insurance brokers company; Industrial Finance Company Limited (IFC); Zambia National Building Society (ZNBS); Zambia State Insurance Corporation Limited (ZSIC); and National Commercial Bank Limited. As the years marched on, at least before the close of the first decade, Findeco seemed to have lost its buoyancy. It was only held afloat by a few viable subsidiaries, such those dealing with insurance and the building society. Findeco's failures dealt major blows to the realization of State goals in the money and finance sector.

Mobilization of financial resources was not the only major problem in this area. The other major problem which drained the energies of policy-makers was administering the funds that were made available to effect the policies. Administering these funds proved not only complex but also caused long delays with the result that some of the programmes fell well behind schedule.

The Transport and Communications Sector

Parastatal activities relating to transport and communications took on some significance with the effects of the UDI. Perhaps the biggest step was the establishment of the National Transport Corporation (NTC), with supervisory powers over the following four subsidiaries: the Zambia-Tanzania Road Services Limited; Bulk Carriers of Zambia Limited; the United Bus Company of Zambia Limited; and Contract Haulage Limited. Some of the companies which made up the NTC Limited complex had formerly been Indeco subsidiaries. For instance, Contract Haulage, United Bus Company and Zambia-Tanzania Road Services. The NTC itself became a ZIMCO Holding Group in 1971.

By and large, the early activities of the United Bus Company form the historical roots of bus service operations in Zambia. UBZ was successor of the Central African Road Service (CARS) when the government acquired 51 per cent holding in CARS after the Mulungushi reforms in 1968. CARS itself was successor to the Thatcher and Hobson Transport Company. These changes resulted in the amalgamation of the bus operations of CARS with those of Barotse Transport Limited. At about the same time, NTC initiated steps which led to the eventual creation of an additional department responsible for operating a Zambia taxi service (Zamcab). The creation of a national taxi service was justified by the need to provide a commuter service in heavily populated Zambian urban areas.

Another important subsidiary of NTC was the Contract Haulage Company. This was created in 1970 from operations carried out earlier by the Zambia Freight Services (a United Transport Overseas Company) and Smith and Youngson, a then Zambian based company. Both companies were engaged in long distance bulk haulage. The merging of the companies and the creation of a single operational control was in keeping with government trends in this area.

The creation and coordination of public corporations in transport and communications helped to maintain and expand the flow of traffic on the country's road network, especially during the UDI crisis period. Other aspects of communications, in particular telecommunications, continued to be run by a quasi-public corporation, the General Post Office, headed by the Post-Master General but, responsible to the Ministry of Transport and Communications for general policy directives.

Other public corporations such as those responsible for the country's railways network or Zambia Airways have not been covered in depth in this chapter. The Tanzania-Zambia Railway has been analysed in previous chapters. All in all, the creation of parastatals in the transport sector was necessary. Unfortunately, their operations were riddled by inefficient management and by haste in the choice of vehicles, some of which were unsuitable for Zambian conditions and their acquisition caused considerable problems to the economy.

The Agriculture Sector

Reference has been made to the central and singular importance of agriculture in developing countries as it relates to economic transformation. Zambia, despite the importance of minerals, was no exception. Parastatals had become one of the country's main elements of economic activity. It was therefore natural that there had to be some parastatals catering for the agriculture sector. The development of agriculture, with the assistance of public corporations, was intended to contribute to an equitable distribution of benefits between the urban and rural sectors and alleviating rising unemployment pressures caused by rural depopulation. It was also intended to use them to help bridge the income gap between the urban and the rural areas.

Agriculture had always been the mainstay of the rural sector. The difficulty in mobilizing effective income generating projects in the rural sector was partly a result of traditional attitudes, intermingled with imported unadaptable concepts of development. Experiments with the creation of several agricultural cooperatives did not succeed. Up to 1968, no less than 900 cooperatives had been established and the bulk of these were agricultural. However, over half that number had proved to be total failure, others were merely lumbering along.

The enactment of Chap. 356, the National Agricultural Marketing Act (No. 30 of 1969), dissolved both the Grain Marketing Board of Zambia (GBM) and the Agricultural·Rural Marketing Board (ARMB). Section 42 of the Act created, in their place, the National Agricultural Marketing Board (Namboard) and Committees thereof. The government was therefore empowered to regulate the functions and duties of Namboard and its Committees. It also had power to regulate and control the prices and marketing of certain agricultural products and agricultural necessities.

Legislation had been used to strengthen operational structure of certain existing parastatals like the Dairy Produce Board, the Cold Storage Board and the Tobacco Board of Zambia. In certain cases, it was used to create new ones like the Rural Development Corporation. This can be illustrated by the range of tasks entrusted to the Rural Development Corporation. It was the legislative instrument which carved up the empire of the Rural Development Corporation with the objective of facilitating cattle financing and farm development which attempted to consolidate and improve facilities on existing farms and ranches. This was to be achieved with funds provided by its subsidiary the Agricultural Development Bank.

However, it is still possible to point to incidences of discrimination in many forms against agriculture. These continued not because of or in spite of the legislation. Some of the incentives given to boost its development failed to make the right impact. While there was no intention to downgrade agriculture, the degree of neglect, partly due to an inflexible bureaucracy, contributed to such things as under-financing and/or inadequate provision of qualified staff. In the same way, the publicly declared policy to give priority to agriculture was rarely followed by the adoption of appropriate strategies compatible with the Zambian environment. A number of projects, especially the ones directed at small farmers in rural areas, tended to be overambitious and too complex to attract the rural population. Instead, small state farms and such institutions as Namboard absorbed the biggest proportion of resources allocated. Yet they proved incapable of generating internal revenues.

Legislation written to support the development of agriculture and institutions such as cooperatives appeared at times to have worked at cross-purposes to each other and the principal objectives. There were several instances when some of the laws to help the development of agriculture simply contributed to nonproductive results. Perhaps the biggest culprit was poor institutional management, coupled with the lack of proper and efficient marketing facilities. Most of these aspects were partly a direct result of rushed legislation.

A major weak point was the lack of provision for coordination between and among many of these institutions. There is very little evidence of collaboration between Namboard and say the Rural Development Cooperation or with the Tobacco Board of Zambia apart from the per chance meeting at diners of colleagues serving on two separate boards. This element was also absent in the operational structures of most of statutory parastatals. There does not seem to have been any attempt made for their collaboration with other developments in the country.

STATE CAPITALISM IN TRANSITION

In every country with no exception, it is expected that governments wish to control or be seen to control the national economy at some stage. In Zambia, there was belief in the early years of adherence to a mixed economy concept. The emergence of public corporations, as indicated in the accompanying chart, was symbolic of a new trend in the economy. It introduced an element of state capitalism in transition. In the Zambian view, weighed against certain government policies, it could be assumed to be an attempt to infuse democratization into the economy. Later popularized as "industrial participatory democracy" by creating opportunities for greater involvement by the nation as a whole.

The Zambian experience in this field was in no way unique. It was in many ways comparable to what happened in other developing countries, especially in neighbouring states. The one important question to be addressed is: Did the Government know the extent to which the economy and its structures were receptive to the new course of action?

The accompanying chart shows an elaborate chain of command in respect of the government's supervisory role of the economy through public corporations. One effort adopted to make the system effective was the establishment of the Ministry of State Participation, but this only lasted a very short period. It combined the functions of several ministries with those performed by Indeco. Also, Cabinet ministers and their respective permanent secretaries were assigned the responsibility of chairmanships and vice-chairmanships respectively, for some of the parastatals

ORGANOGRAM OF STATE PARTICIPATION
April 1971

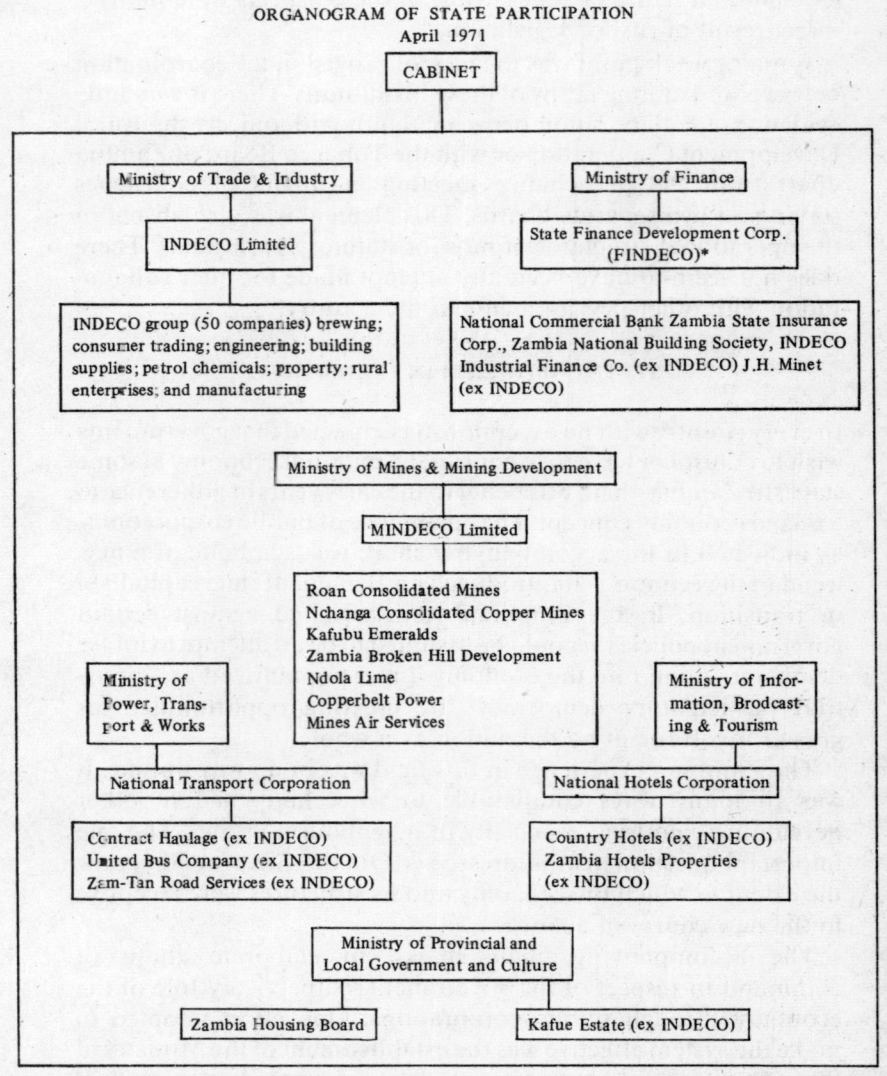

CABINET

Ministry of Trade & Industry
- INDECO Limited
  - INDECO group (50 companies) brewing; consumer trading; engineering; building supplies; petrol chemicals; property; rural enterprises; and manufacturing

Ministry of Finance
- State Finance Development Corp. (FINDECO)*
  - National Commercial Bank, Zambia State Insurance Corp., Zambia National Building Society, INDECO Industrial Finance Co. (ex INDECO) J.H. Minet (ex INDECO)

Ministry of Mines & Mining Development
- MINDECO Limited
  - Roan Consolidated Mines
  - Nchanga Consolidated Copper Mines
  - Kafubu Emeralds
  - Zambia Broken Hill Development
  - Ndola Lime
  - Copperbelt Power
  - Mines Air Services

Ministry of Power, Transport & Works
- National Transport Corporation
  - Contract Haulage (ex INDECO)
  - United Bus Company (ex INDECO)
  - Zam-Tan Road Services (ex INDECO)

Ministry of Information, Broadcasting & Tourism
- National Hotels Corporation
  - Country Hotels (ex INDECO)
  - Zambia Hotels Properties (ex INDECO)

Ministry of Provincial and Local Government and Culture
- Zambia Housing Board
- Kafue Estate (ex INDECO)

*FINDECO would have negotiations succeeded, also have been responsible for supervising and controlling (i) Barclays Bank of Zambia, (ii) Standard Bank of Zambia Limited, and (iii) National Grindlays Bank.

falling within their jurisdiction.

In the meantime, it will suffice only to indicate that to a very large extent, impressive rates of growth witnessed between 1964 and 1976 were a result of massive investment by the state through parastatals. A confirmation of Zambia's style of state capitalism in transition. Under the aegis of these institutions, there was tremendous continuous rise in the ratio of investment to GDP, which was channelled into the public sector. It represented an outlay of K 2,161 million. The efficacy of these manoeuvres is analysed in the next section.

THE LEGAL INADEQUACIES OF USING PARASTATALS FOR DEVELOPMENT

Objectives of the Zambian economic reforms as stated in the previous chapters included a reorganizing of the economy from its former purely capitalist system into a new structure which would not be purely capitalistic. In other words, temper with the old order but without totally eliminating the tenets of market economy structures. However, under the prevailing conditions, the approach had to be selective. The envisaged changes therefore could not have come about overnight. Resort to the machinery of parastatals and the instruments of legislation required a certain period in order to take effect. They were also limited by available necessary complementary factors.

One major influencing factor had been the country's legal system in which parastatals had to operate. By and large, the legal framework was put together in a piecemeal manner. It therefore contributed to legislative interference by some public officials with some of the parastatals.

Professor Harry Hanson is quoted in this connection by Taffara Deguefe on effective control of public enterprises to the effect that "the legistative .... is sometimes over-interested or wrongly interested in the affairs of public enterprises. It has the right to discuss, to advise and to hold the Minister responsible for the exercise of whatever powers he possesses, but it should not be authorised to exercise any direct control over a public enterprise. Such control by a large and amorphous body of non-experts exposes the enterprise to the winds of party politics, makes almost impossible the pursuit by its management of consistent policies and seriously undermines objectivity in the decision-taking process.[10]

The Zambian experience in indigenization, management through the legal machinery and through public corporations reveals that it is easy to promulgate new laws and write amendments but quite another issue to implement changes successfully. Experiences gained in the period under review were both positive and negative.

The shortcomings of certain laws like the Companies Act[11] arose from their having been lifted literally from the old British Companies Act, enacted in the 1920s and partially revised in the 1940s. Transplanting the Act into Zambia and expecting it to fit Zambian conditions without suitable changes was most unrealistic. The least that could have done would have been to update the Act in line with the country's reform programmes, in order to use it as an effective instrument in implementing the country's economic policies.

The Companies Act was weak from the point of view of fostering its objectives. It required every company to register with the Registrar of Companies and Patents and paying a minimum registration duty of K 10 (at the rate of 25 ngwee for every K 200 of business or part of the declared authorized capital of the company). This raises several questions. Was the Act intended to raise revenue, if so then it fell far short of expectations. On the other hand, had it been meant as a measure or scale of investment inflow? This could not be achieved through this Act. Declaring a minimum of K 8,000 as authorised capital for initial registration is immaterial and does not compel the registering company to invest that amount in the economy. Many applicants who had been registered tended to lower the minimum once they had satisfied the Registrar's Office and obtained their certificate of registration. A further weakness of the Act was its lack of policing mechanism to ensure that shares were not devalued to ridiculous proportions.

Figures provided regarding registered companies together with their declared authorised capital over any given period were no measure of the country's economic buoyancy. The register was of limited use as a barometer of the country's growth rate. It was totally inadequate in measuring the extent of indigenization achieved. The directory was useful to the extent that it reflected intentions, not actual operations.

Some expatriate business houses in the immediate post-independence era found it easy to create economic undercurrents to disrupt the economy. Some of their actions were so effective that not even the spate of parastatals was able to save the economy from disaster. For instance, a number of private commercial enterprises who felt threatened resorted to running down their stocks. In other words, they felt less inclined to replenish stocks on their shelves. This action created unnecessary shortages and increased price rises on the black market for many essential goods. There was nothing illegal about such actions because there was no provision in the Companies Act or for that matter in the Industrial Act to prevent them from doing otherwise.

The dominance of expatriates in various economic sectors implied a need for prolonged remedial efforts. Consequently, even some of the expatriate firms which had been taken over were subcontracted, for management purposes, to their former owners. There was very little that the inherited legal system could do to ensure that the retained expatriate managerial personnel acted to further the new national economic order.

The prevailing situation proved above all else, that the State had conferred upon itself only limited power to call the tune. Abrupt stoppage of expatriate operations, initiated because of State action, could have followed by loss of foreign contacts and vital sources of supply. The legal system was incapable of enforcing a continuation of established external supply lines to the new indigenous businesses, whether public or owned by private individuals. Inducements in the form of exemptions aimed at attracting foreign investment did not bring expected results. In the final analysis, they tended to loosen control.

Existing legal provisions were in many cases rather inadequate. For instance, there was no legal provision which could force any expatriate financial or commercial institution to adopt a favourable attitude towards the country's development of parastatals, especially if such a policy did not tally with the wishes of their home base.

The system and the process of economic reforms needed the benefit of sufficiently adaptable laws to enable the government to put the economy on the right road. Revelations of such inadequacies underpin to some extent the contributory factors of the inability of parastatals which relied on the legislative instruments which were themselves uncoordinated. There should at least have been some complementarity among them, so each could help to strengthen the hand of the other.

Many factors prevented rapid implementation of the economic

reform programme. The fact that the Zambian legislative machinery exhibits most of the bureaucratic red-tape of the British system had contributed to the slow progress in this field. Hence, the difficulty encountered with attempting to take arbitrary action or by-pass the various stages of the legislative machinery. The system was the kind which could not be relied upon in matters of pushing quick control over the economy. It created a dilemma between legality and practicability.

Those writing or amending legislation must be aware of the realities of the past and present situations in working towards future development. The legal structure had to balance many interests. For instance, the supervisory role of the State, somehow indicated in the chart. There was also the role of the state as owner and active commercial participant. Similarly, ensuring that the private sector could be made to feel its role and to identify it with the economy in both the short and long run. The legal manoeuvres taken to make parastatals instruments of development were hastily written, inappropriate for the situation and led to inhibiting rather than enhancing efforts to build a strong economy.

REFERENCES

1. Jacques Bugnicourt, "Le Mimetisme administratif en Afrique", *Revenue francaise de science politique*, 1973, p. 1253.
2. W. Friedman, *The Public Corporation*, The Caswell Co. Ltd., Toronto, 1954, p. 541.
3. A. Robson, *Nationalized Industry and Public Ownership*, Allen and Unwin, London, 1960, p. 28.
4. Holis B. Chenery, "Poverty and Progress-choices for the Developing World". *Finance and Development*, Vol. 17, No. 2, June 1980, p. 13.
5. Taffara Deguefe, *A Guide to Service*, Commercial Bank of Ethiopia, 1973, p. 225.
6. Sir Ronald Prain, O.B.E., "Copper: New Thoughts on an Old Theme", addressed to the Mining and Metallurgical Society of America, New York, 26 April 1967.
7. K.D. Kaunda, "Towards Complete Independence", *The Matero Speech*, Zambia Information Services, Lusaka, 1969, p. 28.
8. Ministry of Trade and Industry, "Blueprint for Economic Reforms", Government Printer, Lusaka, 1971, pp. 1-6.
9. A.B. Chikwanda, M.P. "Banquet Speech: The Bank of Zambia New Building", Associated Printers Limited, 1975, pp. 29-31.
10. Tafara Deguefe, *op. cit*, p. 142.
11. Republic of Zambia, The Laws of the Republic of Zambia, Chapter 686— Companies Act, Government Printer, Lusaka.

Chapter Seven

Roads to Development

The plurality of roads to development answers to the specificity of cultural or natural situations; no universal formula exists. Development is endogenous, it springs from the heart of each society, which relies first on its own strength and resources and defines in sovereignty the vision of its future, cooperating with societies sharing its problems and aspirations.[1]

Economic development has never been a spontaneous act. Rather, it comes in stages and must have some beginning. The attainment of political independence by many of the Third World countries represented an historical watershed between different eras. In Zambia, like in other developing countries, it represented one of the several steps of a long march because "even the longest journey begins with the first step".[2] The preceding chapters have revealed several factors about the economies of developing countries, including the fact that the attainment of political independence is not a panacea to solving problems of economic development.

Economists are now in a good position to distinguish states of development and the lack of them in any country. It is generally acknowledged also that underdevelopment or poverty in nations is neither preordained nor immutable. In other words, its elimination can be stimulated or brought about and sustained by deliberate action.[3] The question is to determine the timing and responsibility for initiating the action, and also for ensuring that it is sustained over a reasonable period.

Perhaps we should cite the definition of development put forward in the Dag Hammarskjold Report on the eve of the Seventh Special Session of General Assembly. It was the general opinion that development is whole, an integral, value-loaded, cultural process which encompasses the natural environment, social relations, education, production, consumption and well being of the people themselves.*

In analysing the Zambian experience, an attempt is made to show how some development issues were handled on the basis of the results obtained at the end of the first decade of the country's independence. The question of how successful those measures were is difficult to answer in precise terms. Development activities represented Zambia's first steps after attaining independence.

The analysis has shown that Zambia had similar experiences to those faced by other developing countries. For instance, determining policies and the number of steps or stages necessary to bring about economic growth consonant with economic development. The nature of independence showed that it was far easier to acquire flags and a national anthem than to effect structural changes that lead to a decent life for the people especially in a ten-year time span.[4] People realized that economic growth as represented by high rise buildings or increased volumes of production in only a few sectors, while general incomes remain stratified, could hardly be taken to mean that the country in question was experiencing economic development.[5]

What this boils down to is that the many issues of development can only be tackled by continuous application of adaptable methods. Many inherited and newly generated conditions call for a comprehensive approach and a need to recognize the interrelationship of problems. Writers like Irving S. Friedman refer to this state of affairs as the dynamics of economics and social change. They therefore warn against applying without adaptation, solutions derived from other countries. It follows therefore that an analysis of different parts of the problem is necessary to make possible an overall societal approach, encompassing social, political and economic aspects. Zambia's poverty—including poor nutrition and housing, inadequate education, and a low literacy rate would be more tolerable if there were prospects for gainful and productive employment. Employment problems in developed countries can be regarded as an aspect of national economic management. On the other hand,

---

* It is therefore wrong to assume that governments have a prerogative to develop a country. Rather, they can only help the countries to develop themselves. Basically because "development" is not a condition nor should it be defined by what people possess at a particular moment in time. It is more than anything else, a continuous and gradual process that tends to be cummulative with time.

unemployment in a developing country like Zambia is an extraordinary social problem tied in with poverty and population growth.[6]

What a country does to resolve questions of poverty, employment, and distribution of income will inevitably affect other economic sectors. The dual nature of the Zambian economy causes the rural and urban sectors to greatly affect each other.

On balance, the rate at which a government can accelerate economic growth and development is dependent on natural and human resources available, the performance and structure of trade, political processes, the role of the government itself and its quasi-public corporations and private innovation.

RESOURCES

An obvious problem for orienting and integrating the Zambian economy was the lopsidedness of the country's resources. A monoculture economy existed which was export-oriented but starved of domestic investment. Economic diversification and self-sufficiency under such conditions are then rather difficult to obtain. The snag with export specialization is that some of the country's exports are rarely an absolute necessity to their markets, especially in the industrial countries. Most of those markets not only have their own domestic supply but they also have alternative sources as well. Consequently, fluctuating commodity prices in the 1970s adversely affected domestic policies of developing countries like Zambia.

Zambia's copper reserves, as a percentage of total world reserves in 1976, were assumed at about 6 per cent, or tied for fourth in rank with Zaire (6 per cent) and Peru (6 per cent), after the USA (18 per cent), Chile (18 per cent) and Canada (7 per cent) excluding the proven resources of the Eastern Block, estimated at about (13 per cent). In world production rankings, Zambia again ranked fourth (9 per cent), after the USA (19 per cent), Chile (12 per cent) and Canada (9 per cent) (again excluding the combined output of the Eastern Block countries, whose production was estimated at 18 per cent).

It is possible to estimate Zambia's importance as a supplier, to the EEC for example, on the basis of the above. Zambia's supply share to the EEC copper market ranked second, at 19 per cent, the same as Chile's. Western European countries supplied 6 per cent, the Eastern Block 6 per cent, the USA and South Africa 4.5 per cent each, Australia 3 per cent and Peru 2 per cent. These crude statistical figures do not reflect other important aspects such as each country's production and shipping costs, and the role of transnational corporations in determining supply.

Zambia's share of total world cobalt reserves is 8 per cent, after Zaire 31 per cent,. New Caledonia 18 per cent and the Philippines 13 per cent (excluding the share of the Eastern Block 21 per cent). Zambia's world production share was 9 per cent second after Zaire's 53 per cent. The other important producers of cobalt are the Eastern Block 19 per cent, Australia 7 per cent, Canada, Morocco and New Caledonia at 5 per cent each and Finland 4 per cent. The EEC's most important source of unwrought cobalt metals was Zambia 33 per cent, Zaire 24 per cent, the USA 19 per cent and Norway and South Africa 7 per cent. It goes without saying therefore that Zambia is in no position to influence the prices of her major export minerals on the world market, apart from negative domestic actions, like strikes. Strikes negate the country's ability to earn the adequate foreign exchange needed to finance development projects.

This overreliance on the export of raw mineral resources and the existence of an underdeveloped agricultural sector contributed to some of the country's economic development dilemmas, especially in the face of "an inelastic" world demand (a situation in which price fluctuations exceed change in quantity). When the monoculture syndrome operates in a dual-economy, such as Zambia's, it leads to a highly productive export business which is segregated from the domestic sector. Mining has benefited the economy by providing the country with the bulk of the foreign exchange revenues used in sustaining growth, trading and paying for deficits. Unfortunately, development of the domestic sector has been negatively affected.

On the basis of a number of generally accepted economic indicators, Zambia is among the so called *middle-income* economy countries, with a GNP per capita of $ 560 in 1980. However, the country has difficulties in trying to expand its exports and has only a very modest manufacturing capacity.

It is possible to understand some of the development difficulties faced by the Zambian economy from data given and the analysis in chapter one. The tables are sufficiently self-explanatory. Not

TABLE 7.1: Zambia's Major Exports*
1967-1970 and 1973-1976

	1967	1968	1969	1970	1973	1974	1975	1976
(a) Value (US$'000)								
Total	654,000	757,000	1,056,000	995,000	1,136,229	1,399,351	805,078	1,043,800
Cotton	519	179	12	—	2,600	2,700	3,700	—
Maize	12,250	3,973	524	—	4,069	11,861	2,228	710
Wood & timber	839	893	933	741	773	—	—	—
Tobacco	5,178	3,704	4,428	3,993	7,355	9,018	7,722	7,045
Cobalt metal	7,870	4,724	6,350	8,879	7,419	12,317	10,980	22,077
Copper	607,673	720,860	1,014,300	952,178	9,081,017	1,294,854	721,249	935,129
Lead metal	3,775	3,916	8,485	6,825	8,370	11,112	8,803	6,049
Manganese	732	559	—	—	—	—	—	—
Zinc metal	11,294	12,546	17,364	15,345	25,906	39,103	31,619	33,722
(b) Quantity (metric tons)								
Cotton	900	278	21	—	4,300	3,300	3,900	—
Maize	198,097	64,000	8,442	—	50,086	111,212	16,621	8,808
Tobacco	4,449	3,629	3,768	4,041	5,048	4,872	5,337	4,751
Cobalt metal	2,091	1,227	1,588	1,814	1,145	1,894	1,344	1,976
Lead metal	17,413	16,637	25,891	22,082	20,012	18,776	19,376	14,610
Copper	610,000	641,158	730,136	682,403	671,000	763,400	639,760	746,100
Manganese	24,710	17,483	NIL	NIL	NIL	NIL	NIL	NIL
Zinc metal	39,804	45,026	53,886	50,343	51,115	50,227	41,264	48,941

*F.O.B.—Preliminary: *Monthly Digest of Statistics*, Nos. 7 to 9. 1978. Republic of Zambia (Converted to US $ by factor L26753).

Source: *Monthly Digest of Statistics*, Vol. X. No. 6, Central Statistical Office, Lusaka. Zambia, June 1974; *1978 Yearbook of International Trade Statistics*, Vol. 1, United Nations, New York, 1979; ECA Statistics Division

TABLE 7.2: Zambia's Major Imports
1967-1970 and 1973-1976*

	1967	1968	1969	1970	1973	1974	1975	1976
Total	423,650	504,640	435,000	477,000	532,044	787,544	930,132	674,000
Food and live animals	31,000	34,000	43,000	43,000	37,342	68,070	56,554	36,000
Beverages and tobacco	2,000	3,000	2,000	2,000	1,492	2,000	2,000	1,000
Crude materials	6,000	6,000	6,000	7,000	9,314	16,106	15,372	10,000
Mineral fuels	44,000	46,000	50,000	49,000	51,056	94,947	126,061	101,000
Animal, vegetable oil and fats	3,000	3,000	4,000	6,000	6,631	10,521	14,121	15,000
Chemicals	29,000	32,000	32,000	36,000	53,552	74,825	119,567	95,000
Basic manufactures	93,000	104,000	89,000	105,000	119,101	203,547	218,901	135,000
Machinery and Transportation	176,000	188,000	172,000	184,000	213,440	258,028	328,662	233,000
Miscellaneous Manufactured articles	39,000	34,000	35,000	43,000	37,999	54,998	43,704	27,000
Miscellaneous transactions	5,000	6,000	2,000	2,000	3,117	4,502	5,190	1,000

Source: 1970 Yearbook of International Trade Statistics, Vol. 1, United Nations, New York, 1973 ECA Statistics Division, 1980.
*F.O.B.

only do they explain the main sources of foreign exchange earnings but they also relate the earnings to the country's import spectrum. For instance, earnings derived from exporting manganese ores dropped out of the picture after 1968, while those from exporting wood and timber declined after 1969, followed by considerable fluctuations in maize exports. On the other hand, the country's imports increased throughout the period.

This apart, were other questions, such as how to ensure that inflationary tendencies did not accompany deficit financing in the investment process. Tables 7.1 and 7.2 show that imports exceeded exports by a very big margin. These aspects assumed greater importance because the bulk of the country's development revenues came from the copper industry and taxes paid by the big, non-mining companies.

In the period under review there was a disproportionate increase in recurrent expenditure, caused in part by the high rate of capital investment. The high earnings from exporting copper, gained immediately after independence provided grounds for optimism and gave encouragement to the government in drawing up many investment projects. However, fluctuations in world mineral prices considerably reduced the replenishment which was necessary to maintain the momentum of economic growth and development.

The problem of resource constraints was complicated further by the external environment, especially with the UDI and the spillover effects on the Zambian economy, as analysed in Chapter Two. Zambia had been able to increase her exports in 1970, but efforts to significantly reduce imports, especially after 1975, were not as successful. The country's inability to reduce the range and volume of imports was a major option problem and contributed in part, to large borrowing of commercial capital. The development options that emerged tended to be characterized by a high consumption rate and which were much seriously strained by the need of providing services at a higher rate of over 18 per cent of total income.

At the macro-economic level, the structure of production continued to be grossly dominated by mining, which accounted for over three-quarters of total domestic product. The industrial base continued to be relatively small and made more use of imported rather than local inputs into production. It was also fractured and only minimally linked to the country's natural resources base. In addition to such internal structural imbalances, the economy remained the most exposed to the world at large because its external trade assumed an upper hand.

DIRECT GOVERNMENT SECTORAL LEVERAGE

We have analysed direct government intervention in the economy through the establishment of parastatals where the principal concern was increasing production and income distribution. Zambia, in line with other developing countries, wanted to create a system which could help accelerate the pace at which the nation could change the dual nature of its economy.

In consequence, many of the quasi-public institutions established in Zambia under the aegis of Zimco were mobilized in pursuit of this goal, using a number of roads to development. Other developing countries such as Israel, Yugoslavia, Cuba, the People's Republic of China, India and Sri Lanka have, in the recent past, attempted to attain both growth and alleviation of poverty. In the Zambian context, this was to be provided by the FNDP and SNDP as revealed in the analysis in Chapters Three and Six, by dispersion of government revenues to the various sectors.

Zambia's experience with economic activities is illustrative of some of the problems which could only be solved by direct government action. It was quite obvious that there were a number of other aspects of economic activity in which government influence was difficult to exert, despite the promulgation of legislation.

For instance, the need to increase managerial expertise in several parastatals, in order to efficiently run the economy. This was one function which legislation, no matter how enlightened, could not produce in a short period of time. There was a very small pool of trained indigenous persons with skills which the government could draw on to maintain both the government and quasi-government, machinery, in addition to fulfilling the requirements of the private sector.

Attempts to augment education facilities contributed to the confusion. This was due partly to the lack of defined territorial responsibilities in the field between education and training. The

Ministry of Education, on one hand and the Vocational and Technical Training arm of the Department of Manpower Training on the other, seemed to fight for territorial supremacy. There was also the tendency to confuse the provision of general academic education with creating managerial capabilities. The fact that an educated person is capable of grasping and learning managerial skills more readily than his uneducated counterpart did not necessarily equip the former for senior management posts. Table 7.3 in this chapter helps to show the inappropriateness of some of the appointments made in the country's state trading organizations. Management for these institutions should have been understood to aim at serving, identifying and improving the people's standard of living. This should have been one of the criteria used in selecting those who were appointed to head the various government commercial enterprises. However, evidence shows that this was not always the case.

Table 7.3 also illustrates Zambia's levels of skills in 1973 in most of the country's state trading organizations. It is collaborated by a similar investigation conducted, during the same period, by Walter Halset of the International Labour Organization. Halset found evidence of personnel deficiency at most levels. For instance, several supervisory managers did not possess the necessary managerial training and experience for the posts they were in, when assessed in terms of their training to improve their skills. The inappropriateness of such individuals as reliable tools in direct government efforts at sectoral leverage of the economy is obvious.

Along with the problem of trying to create a pool of managerial skills were issues of using direct government intervention in resolving unemployment. Creating job opportunities was a task not only for the current but also for future generations. Varying estimates of unemployment, both in the urban and in the rural sectors, were analysed as part of the 1966 and 1969 population census programmes. The unemployed were categorized and listed as follows: the so-called low income groups; frustrated job-seekers; the under-utilized or inappropriately employed; and womenfolk engaged in unpaid domestic jobs.

The government's attempt to alleviate unemployment laid stress on massive investment in public works and in quasi-public corporations. Consequently, no less than 30 per cent of the GDP between 1965 and 1974 was invested in the economy with the aim of increasing employment opportunities. Translated into monetary terms, it represented a capital injection of approximately K 3,500 million. Different methods were adopted to address the problem. In particular, there were differences between handling the urban and rural umemployed.

Most of the capital investment earmarked for employment creation in the rural areas failed to achieve their objectives because of the use of inappropriate techniques. Most of the techniques used in the rural areas were not labour-intensive and were run by large firms which relied on large-scale production units and the use of highly mechanised capital intensive machinery. Another depressing feature of public policy, using the SNDP, a reflection of the gulf between words and deeds, was that only a mere 7.3 per cent or K 152.5 million of total capital investment, was allocated to the rural sector.

Sometimes choice of a particular course of action was unavoidable. For instance, choosing between preserving or losing a crop, particularly in a situation where available indigenous labour could not be persuaded by existing labour incentives to take up job opportunities in the remote rural areas. The negative attitude of urban Zambians in instances of this kind was prevalent and cast a shadow of doubt on the future of a number of state farms. Often, able-bodied male unemployed Zambians turned down farm jobs in the remote rural areas in order to remain in urban areas.

Consequently, the government resorted to capital intensive methods of farming even in the face of great national unemployment. Thus, the harvest of the cotton crop at the Chombwa scheme had to be mechanized rather than rely on the country's reluctant human labour force.

The rather disappointing results with integrating the rural sector into the economy was in marked contrast to government results in fostering the process of self-reliant dynamic growth and diversification through the establishment of parastatals. Chapter Six discusses the emergence of several public corporations, which assumed the dual role of planners and performers in many sectors of the economy.

Economic analysts no longer use GDP as the only major measure of economic buoyance. This is because they generally acknowledge that GDP trends only provide a measure of overall

Roads to Development

growth of the economy. The Zambian economy, in terms of GDP at current prices, more than doubled in 1973 as compared to 1965. It increased from K 650 million to about K 1,352 million, representing an annual growth rate of about 9.6 per cent or in real terms, about 4.0 per cent. This increase took place in spite of variations in the period under review. For instance, during the period 1965 to 1969, the GDP growth rate was as high as 17.5 per cent, but was followed by a decline, during 1970 to 1972, and then a fairly substantial rise of about 11.2 per cent in 1973. Corresponding growth rates in real terms for the period 1965 to 1969 were 11.1 per cent, a decline during 1970 to 1972 to about 4.2 per cent and an increase in 1973 to 8.5 per cent.

One of the most significant factors contributing to the above fluctuations was the Mufulira Mine cave-in. The disaster could not have come at a worse time in Zambia's economic history. The Mufulira Mine was one of the biggest copper mines in the country and it took a very long time to put it back into production. External factors of great significance included the international monetary uncertainties, the closure of the Rhodesian border with Zambia and the instability of metal prices. The fact that the country was able to reap certain benefits for its economy as a result of an increase in world copper prices in 1973 was fortuitous.

Zambia still bore the scars of economic exploitation suffered during the unholy wedlock of the Federation of Rhodesia and Nyasaland. Overtures for closer economic cooperation with countries of the former East African Community on one hand, and Botswana, Zaire or Malawi on the other, had been tried but were not finalized. Consequently, the emergence of new industries even if fashioned on the lines recommended by Raul Prebisch for the sister countries of Latin America, i.e., the utilization of some of the foreign exchange proceeds derived from a country's raw materials exports, had to be adopted, taking account of the above constraints. The lack of indigenous skills made this task rather difficult.

The Zambian road to industrialization could not follow the patterns of other developing countries, especially those in the Latin America region where, as early as 1950, Raul Prebisch had an import substitution recipe. The folly of import substitution as a panacea for economic ills was amply demonstrated from experiences in that region. In Zambia, Government investment

leverage had to deal not only with supply problems but also with a production and distribution system country-wide.

Initiative in industry in Africa, and in Zambia in particular, is not a common trait. The Zambian experience confirms that it was hindered by traditional concepts. The traditional concept of an African entrepreneur assumes that motivation to produce goods in return for some benefits as well as using managerial capability to identify inputs and organize the production and marketing of final products. The value of pre-feasibility and feasibility studies tends not to be appreciated.

This situation could have been overcome by creating national consultant engineering companies. They could prepare project designs for potential indigenous entrepreneurs and enterprises. Government policy should also have demanded that all large consulting contracts awarded to foreign management enterprises and consultant engineers should provide employment for indigenous graduates. Somehow this would have facilitated the transfer of technical know-how and skills.

The need to increase indigenous technical and managerial capabilities is essential in all developing countries of the Third World. Gaining that increase is governed by a number of goals, motivations and organizational behaviour. It is therefore the responsibility of governments to implement policies and strategies aimed at achieving the goal and the social objectives of self-reliance and self-sustaining development. The United Nations in its "Programme for the Industrial Development Decade in Africa" (a joint publication of UNIDO, ECA and the OAU (ID/287 at page 40) also advocates a similar point of view and goes on to call for high priority to be accorded to establishing effective industrial extension services, to provide financial, technical and marketing advice to small-scale, handicraft and cottage enterprises, in order to contribute to effective rural development.

The spread of industrial activities in Zambia between 1964 and 1973, particularly in manufacturing, increased by about 31 per cent. The number of units rose from 253 on the eve of independence to about 550 by October 1973. Many of these were established with direct government support. Impressive as the numbers may be, subtraction effects must be considered. The numerical presentation was simply a head-count on the basis of the register provided through the Companies Act, which we discredited as not a reliable measure of industrial activity.

This list of notable agro-industries established during this period included the pineapple scheme, with its associated canning factory in the North-Western Province and the extensive sugar production scheme at Nakambala, near Mazabuka along with various refineries. The establishment of other industrial projects included the copper fabrication factory, the provision of a network of electricity supply, using local hydro power, glass bottle manufacturing and several assembly plants.

The development and expansion of the country's sugar industry was a major industrialization effort. However, sizable amounts of foreign exchange had to be made available and this was an important factor influencing the government's acceptance of non-Zambian commercial partners like Tate and Lyle in the sugar industry. Benefits in sugar refining included the increased use of local materials, and its direct relationship with other industries. For example, a number of products can be produced in combination with sugar as an important ingredient. Its by-products are an essential ingredient in producing a variety of animal feeds, alcohol, bags and some paper products. Sugar itself is a base for confections. Nonetheless, the industry was plagued with maldistribution from the refinery country-wise, necessitating remedial measures in order to improve its profitability.

The establishment of the Zambia Metal Fabricators Limited (Zamefa) was a major industrial breakthrough during the first decade of Zambia's independence. It introduced a significant change by showing Zambia's capability to export its copper products in a refined form. Thereafter, some refined copper was manufactured locally and used to produce items intended both for the domestic and export markets. This particular government initiative increased the range of the country's exportable items with the inclusion of copper rods, wires and cables. The country also witnessed the establishment, by the private sector, of factories on the copperbelt which produced sophisticated copper-craft wares.

The role of the government in fostering economic development by direct intervention included providing infrastructure. Communications by land routes, in association with air and telecommunications, needed tremendous improvement during this period. This was an area where the private sector would be less inclined to invest its resources given the rather long period for recovery of the investment. The main railway line linking Livingstone, Lusaka and Kabwe to the Copperbelt was joined at Kapiri Mposhi to the new Tanzania-Zambia Railway (Tazara) only because of direct government intervention and investment. A major internal railway development was the construction by the government of an additional coal carrying spur from Choma to Masuku. Zambia's total length of railway route increased to over 1,900 kilometers by the turn of the first decade of independence.

Steps were taken to provide the country with more roads. About K 109.5 million was earmarked for road construction and maintenance between 1964-65 and 1973. The total length of all classes of roads rose from 33,316 kilometers to 34,671 kilometers between 1964 and 1973. Paved roads in 1973 accounted for an approximate total length of about 3,220 kilometers out of that total. This record may seem less impressive. However, it must be weighted against some of the odds which absorbed the country's energies and resources in the period under review. In particular, rerouting and providing alternate transit routes as a result of the UDI.

The country also established and implemented a long-term aviation policy. Aircraft movement on the domestic level makes it possible for better and quicker links and contacts between places. Air routes opened up the country's great horticultural potential. In terms of external activities the country's being landlocked, it was air communication routes which saved Zambia from economic strangulation during the early years of the UDI when the southern border with Rhodesia was closed unilaterally by the Rhodesian regime.

The government also invested heavily in the other infrastructure development, including teleprinters, telegraphs, radio, telephone and television. Roads, railways, and air traffic do not function smoothly in the absence of an efficient telecommunications network. This particular sector had been greatly neglected during the colonial era. Many districts, especially those not along the line-of-rail, could not be easily reached by telephone or telegraph, let alone teleprinter services. Many places could only be reached on specified and restricted times or days.

Electricity supplied to consumers more than doubled between 1964 and 1974. The generating capacity increased four times during the same period. Its provision over a much wider area of the country raised possibilities for greater development. Considerable inroads were made in reducing the country's dependence on supplies from Kariba South, which had always been under threat of closure by Rhodesia.

It is good that Zambia had the foresight to develop her hydro-electric power. The development of more electric power is partly what saved the Zambian economy from collapse after 1973, when world oil prices soared to astronomical levels. The pattern of energy consumption, especially oil fuel before 1973, could not be sustained by the economy over a long period given the new policies of transforming industry and agriculture. High oil prices had forced Zambia to adopt economical ways in which to use her energy resources. The adjustment process which followed would have been very difficult had the country not developed its hydro-electric power stations. Some of the power from the country's hydro-power stations brought in sizable foreign exchange revenue from Zaire and Southern Rhodesia.

Direct government leverage in each of the above sectors was necessary, brought about positive results and laid the foundation for future development projects. For instance, changes in agricultural and industrial policies encouraged new urban and certain rural centres (such as the institution of Intensive Development Zones, Village Productivity and Ward Development Committees) in agricultural areas, with strong financial and fiscal power making the management of future urban and environment problems much more possible. However, these things were not pursued with the needed determination during the first decade of Zambia's independence. Prior to the construction and completion of a network of telephone or telegraph, let alone teleprinter services, many places could only be reached on specified and restricted times or days.

Considerable effort was therefore exerted, through direct government action, toward providing full telephone and telegraph services to almost all districts during the first decade of independence. In particular, the high grade trunk routes were installed serving all provinces with sufficient capacity to maintain anticipated future demand. The overall system also took account of the projected Pan-African Telecommunications Network (PANTEL), designed to link Zambia to her East African neigh-

have made it possible for more people to live comfortably on less land.

The extent to which land is useful to Zambia's economic future goes beyond producing minerals and primary agriculture products. Unused and exhausted mines might serve as towns and cities of the future to accommodate the population explosion when we consider the current practice of building bomb shelters and similar bunkers. The country's population by the year 4000 could well force a thorough exploration of this rather remote option.

There is considerable room for agricultural expansion. Many of the initiatives of the first decade of independence were taken to achieve this very objective. Zambia's main constraints consisted of a lack of agricultural entrepreneurial skills, capital, access to markets and the institutional inadequacy in dealing with credit and marketing. The first of these constraints is a legacy of colonialism. Agricultural as well as commercial skills take many years to develop, even with the help of agricultural extension services and rural development programmes.

What is the optimal use of Zambia's agricultural land? As if in answer to the above, the country proceeded to mobilize the large commercial farmers. Attention should also have been turned to the African small holders. There is a lot of evidence, in Zambia as elsewhere, that these people respond rationally to economic incentives, notwithstanding the success time lag. Conservatism on their part stems in part from their poverty. Change, any change, involves an element of risk. It is therefore not unreasonable for a poor people to avoid risk. There have also been incidences where pricing policies adopted by certain state marketing institutions, such as Namboard of Zambia, have retarded expansion and distorted benefits. Similar incidences have been analysed by Peter Temu in his study of "Marketing Board Pricing and Storage Policy" in Tanzania and Kenya.[7]

This analysis reviews data on land-use in Zambia by 1976, as derived from the 1977 FAO production Yearbook. It provides knowledge of how the country performed in the agriculture sector in the study period. Of the total land area of 74.1 million hectares, about 5 million hectares (6.75 per cent) was used as arable land for permanent crops. Another 30 million hectares (40.5 per cent) was used for permanent pastures (i.e., land used for five years or more for herbaceous forage crop, both cultivated and wild), while forest and woodland occupied about 37.1 million hectares (50.3 per cent). There was a wide range of crops grown on the arable land, made possible by irrigation schemes. The list of crops grown included the traditional ones, maize, millet, cassava, sorghum, sweet potatoes, groundnuts, bananas and tobacco, and others, such as cotton, oil seeds, citrus, coffee, rice and wheat.

The performance of the agricultural sector in the SNDP was below its target rate of growth of 5 to 6 per cent but much better than that achieved during the FNDP period. The GDP from the agricultural sector (including forest and fisheries) between 1965 and 1970 and between 1971 and 1976 grew at an annual rate of 1.8 per cent and 3.6 per cent respectively. Like most statistical data, the above information conceals other significant developments, especially in the crop sector. By taking 1971 as base, it will be observed that the economy had achieved the set target of the SNDP. For example, the target of 8.5 per cent annual growth rate for marketed production in agriculture was exceeded in respect of maize and sugar. Noteworthy also was the fact that it was during the SNDP that the country harvested its first crop of wheat. Other significant achievements were in respect of increased beef production except for the fact that a low target had been set for the SNDP.

However, the situation remained serious in respect of certain other crops. For instance, augmenting other crops like vegetable oils (groundnuts) by growing sunflower, increasing the production of tobacco seed cotton and dairy products had not been successful. The country had failed to attain the target set in the SNDP.

SUBSIDIES

Subsidies are used to integrate an economy and in fostering a balance among the various sectors. Subsidies can be considered as a way of tackling inflationary pressures in an economy. However, in Zambia, they were used to prime a number of ailing public institutions which could not survive commercially. Consequently, the government had to admit in 1967 that subsidies were no longer acting as anticipated. Their folly and the limitations of public enterprises that continued to make losses showed that "subsidies of any type, whether on maize or any other commodity, not only result in a misdirection of resources, but were also apt to put a premium on: inefficiency unless they are given after careful scrutiny of the various factors going into the cost of production of commodities concerned, as well as the overall working of the production units concerned."[8]

In the period between 1965 and 1975, were a number of thorny issues as to the actual beneficiaries of the government's subsidy policy. There is very little evidence to support the view of priority having been accorded to the rural sector and those industries which were to increase their use of local resources in their production process. Unfortunately, many subsidies consisted of payments to compensate for price increase and for the losses incurred by public enterprises. There was therefore a notable sharp increase in the outlay for subsidies. It rose from K 4 million in 1965 to K 16 million in 1966, to K 24.1 million in 1967 and to K 34 million by 1974.

Developing countries especially those with dual economic structure cannot easily disperse with the "carrot" of subsidies. The non-monetized sector of the economy was dominated by subsistence characteristics. It was unlikely to maintain itself without direct government support in the form of subsidies, not only towards the essential foodstuffs but to other communal infrastructures such as communication, health and education, Consequently, the government saw in them an element of fiscal and monetary policy in priming the economy, as was made abundantly clear in the 1974 budget speech.

An analysis of the impact of subsidies in an economy includes answering the question—what enterprises should a government subsidise? Joram Mayshar attempted to analyse the reasons behind government subsidies to private projects.[9] Nachum Finger took the Israeli experience to show the impact of government subsidies on industrial management.[10] Both examples illustrate the problem of resource scarcity as related to subsidies.

Zambian examples of misdirected subsidies are the Zambia Taxi Services (Zamcab) which was partly fashioned on lines similar to the Kenya Transport Company (Kenatco taxis). Zambia's failure to mature as a viable commercial enterprise was partly administrative. The example also serves to underline the inability of parastatals in certain sectors of a country's economy. The company was riddled by problems of mismanagement.

Drivers employed by the company also tended not to turn in the full days's fares. This was contributed to by a lackadaisical attitude of the supervisors, most of whose appointment was very questionable. In addition was the inappropriate vehicles used in the fleet and destroyed by employing incompetent drivers.

Subsidies were also used to establish and to prop up many of the country's cooperative societies. Proponents of cooperative societies often pointed to their being effective instruments in preventing the degeneration of the Zambian society. They thought society was becoming adulterated by material considerations, in disregard of humanist teachings on the importance of man. Thus, it was assumed that continuation of subsidies was the price to pay, even in the face of certain economic disasters. It was more important to revitalize Zambia's traditional communal life, which embraces ideas of communal accommodation, interdependence and reciprocity.[11]

Many cooperatives set up in Zambia were not able to meet their principal objectives. Some of them simply complicated and aggravated the dilemmas of economic development and wasted away limited resources. It is worth recalling a few examples of the actual working of some cooperative ventures. M. Lungu, at a seminar sponsored by the Scandinavian Institute of African Studies, in 1978, cited how the Shangombo Farming Cooperative Society Limited and the Mumbwa Building Cooperative were riddled with gross mismanagement and corruption.[12] Many cooperatives were led by groups of misguided self-assuming office bearers, who under normal circumstances, would not have been permitted to run such enterprises. There was also a lack of proper supervision by the central government authorities.

The intention is not to condemn the establishment of cooperative societies. But the Zambian experience, like that of Sweden or Israel, has to be set in the right perspective as far as the element of subsidies is concerned. The fact that a greater proportion of the Zambian population lives in the rural areas and derives their subsistence therefrom ought not to have been the only sufficient economic reason for subsidizing cooperative societies on a country-wide basis.

There was a strong urge to assist with transformation of the rural sector, as well as narrow the gap between it and the urban sector. Also at the back of the minds of those who advocated the creation of cooperative societies was the modernization of the rural economy without breaking down community life and traditional social ties. A factor on which Father Bede Onnoha placed greater emphasis in his analysis of the Nigerian situation.[13]

It therefore becomes clear on the basis of the above that Zambia's economic renaissance was grappling with many difficulties, principal among which was how to increase capital accumulation savings. The subsidy element is therefore not a separate element from the propensity to save or accumulate capital reserves. In view of the country's limited traditional foreign exchange earnings from mineral extracting and manufacturing industries, trade and banking firms, the problems were obviously great.

SHARING DEVELOPMENT RESPONSIBILITY

Development has been defined in the preceding chapters to include increasing the output of goods and services available to the community, creating structural change in production system and preventing social consequences of uneven distribution of income as well as developing responsibilities of the state. It is a task which cannot be undertaken by an individual or one sector in the economy. It therefore has to be a shared responsibility of all individuals and sectors. There are several ways and forms of sharing development responsibility at both national and international levels.

The above conceptual framework provides a basis for the trends observed in post-independence Zambia. At independence the new Zambian government took over the reins of political control while control of the economy remained entirely in the hands of non-citizens. The indigenous populace had no genuine commercial and economic tutelage or opportunities. Naturally, no responsible government would allow a situation of that kind to continue. Consequently, greater economic participation by the indigenous community became a major goal. The government had to induce an atmosphere in which the private industry, both foreign and local, would cooperatively share in the commercial and industrial life.

Parastatals were instrumental in accelerating economic growth and development. As a resut of their work, the government found it relatively easy to acquire the ability to formulate and execute, through them, its economic and social policies.

From the government point of view, the creation of a battery of parastatal organizations was supposed to represent a strengthening of the development process. On the contrary, some of the failures exemplified by the activities of those like Namboard, could only retard progress. There was a certain degree of loss of control by the government. Late in the day, the central authorities realised that they were only creating a formidable competitor in the power arena rather than sharing development responsibilities. This resulted from a majority of the parastatals being more concerned with their internal growth than with development activities. On balance however, the emergence of parastatals was a mixed blessing for the economy as a whole, even if it did not give much direct effective control to the indigenous population.

Some other difficulties contributed to the lack of shared responsibilities among the various sectors despite the principal objectives enunciated in the reform programmes. For instance, there had not been a tremendous increase of Zambian participation in all sectors of the economy. Efforts to bring about this change encountered several obstacles. Some of the measures taken to cope with the situation included restricting expatriates from borrowing from Zambian banks. Thus withholding from them retail and wholesale activities were more a mere bother than they were worth. Actions of this kind did not really increase indigenous participation. As a matter of fact, most measures were aborted by the newly created parastatal structure. For instance, a number of Zimco subsidiaries acquired wider trading activities without sparing a thought as to their implications on individual indigenous Zambian entrepreneurs. The latter were disadvantaged because they were incapable of competing with the former's machinery which was buttressed by government financial support.

The expected sharing of responsibilities with indigenous individuals was greatly limited. The only areas where this had meaningfully taken place was in respect of the parastatals with the new government. In spite of this inroad, many public institutions suffered from a conflict of policy objectives. Some were expected to wear two hats at one and the same time and strike a balance between two seemingly opposing objectives. For example, the ob-

TABLE 7.3: Synopsis of Parastatal Supervisory Posts (1973)

	Total and Sub-total	Deficiency of qualifications	%	With partial required qualifications	%
(i) Managing Directors and General Managers	579	116	20.0	463	80.0
(ii) Managerial posts (minimum Form V and/or 3 years Diploma)	5 000	2 900	58.0	22 100	42.0
(iii) Managerial posts	2 400	850	35.4	1 550	64.6
(iv) Deputy General Managers Marketing Managers Sales Managers	632	253	40.0	379	60.0
(v) Managers (Administration)	632	259	41.0	373	59.0
(vi) Assistant Managers	634	252	39.9	391	61.9

Source: Compiled from data in *Zimco Annual Reports to 1976* and the *Record of Zambian Graduates in Government Service, the private Sector and Quasi-Government Institutions,* Directorate of Civil Service Training. Republic of Zambia, 1 September 1972, pp. 29-43.

jectives of operating as viable commercial enterprises on the one hand and providing a national social service on the other. The two objectives need not be mutually exclusive. However, such tasks could not be achieved with the sort of tools then at the disposal of many Zambian corporations. The National Transport Corporation (NTC) and Namboard could have coped with these challenges if only they had enjoyed autonomy and independence in their actions and management. However, for some corporations, total independence would certainly have led to disaster given the low levels of management skills, revealed in Table 7.3.

Some of these mismanagement problems came to the fore in the case of Namboard's impact on the country's agricultural development. An observation made by one of the nation's dailies illustrates the problem: "in 1965, this country (Zambia) destroyed the dairy industry because we arbitrarily decided to bring down the price of milk to margins where dairy farmers had no alternative but to call it a day.... And in 1970-71 season, the maize crop was so bad that we had to import maize ... at a cost of K 33 million. All this was because farmers had decided to turn to cash crops after drastic cuts in the price of maize to farmers.... Last year (1976), the cattle industry faced a severe jolt when cattle and beef prices were arbitrarily slashed down without consulting the cattle producers.... On Saturday, last week (7 May 1977) ... the National Agricultural Marketing Board were reported by Zambia News Agency (ZANA) to have introduced new potato prices.... What these prices mean is exactly what ZANA had reported which is that the prices of potatoes in the country had come down by 50% .... Only four years ago, there was a hue and cry in this country when we learnt we were importing potatoes from Rhodesia, South Africa, and as far afield as Australia".[14]

On the other hand also were situations exemplified by activities of such institutions as the Development Bank of Zambia (DBZ). The Bank (DBZ) was created to stimulate capital expansion, assist the Government in allocating capital funds for investment in high priority industrial sectors and to mobilize external financial resources for certain categories of industries. It was therefore of strategic importance to national economic development. The Board of Directors of DBZ tried to escape pitfalls of its predecessors, like Land Bank, IFC, COZ by avoiding indiscriminate loan giving. Within the DBZ itself was a system for supervising the way its loan recipients used the money lent to them. The Board also demonstrated a clear sense of economic maturity by rejecting the notion that business take-overs constituted actions of an economic development nature. Thus, applicants in this category were accorded a much lower priority in resource deployment.

The operations of the DBZ included assistance to private firms in their initial growth stages. It also rescued ailing industries. In other words, it helped stave-off economic stagnation, especially where there was danger of swelling the rate of unemployment. However, DBZ took far too long to resolve the problem of distinguishing between bankable and non-bankable projects, on both priority and non-priority basis.

There was an awareness on the part of all sectors that development of the national economy could only be achieved as a shared responsibility. Perhaps this is what made possible the emergence of a fair number of medium- to large-scale indigenous Zambian enterprises. A majority of the Zambian entrepreneurs were able to succeed through a process of take-overs with funds provided through Indeco, Findeco subsidiaries and/or the

Development Bank of Zambia (DBZ). Men like Emmanuel G. Kasonde were able to start completely new enterprises, manufacturing such items as flexible packing materials, cellophane wrappers, waxed and unwaxed paper wrappers, poly bags, poly tubes and poly film. Kasonde was also singularly enterprising in having been able to obtain financial support from as far afield as the International Development Association (IDA) for some of his business ventures. The Chibote (Henry Shikopa and Andrew Sardanis) conglomerate, with interest in a wide range of activities, was yet one other successful venture by indigenous Zambians.

If ever there were Zambian successful examples of individual entrepreneurial zeal, then that of Flavia Musakanya should be correctly reflected. She against considerable odds had managed to make inroads in the area of agriculture by exporting flowers, raising as well as exporting dressed guinea fowl meat. These exports not only earned Zambia a lot of foreign exchange, but also showed that with better industrial and agriculture policies or climate, Zambia's economic potential had as yet to be fully exploited.

Nonetheless, some other detrimental developments ought also to be listed. There were certain other aspects which tended to discourage rather than encourage indigenous private initiative. A brief mention of the introduction of the Leadership Code is one such example of a possible discouraging aspect for certain sectors of the economy.

The Leadership Code was adopted in December 1972 in the context of the Kabwe Declaration by the United National Independence Party (UNIP). The philosophy behind the introduction of a Leadership Code was to provide guidelines for the economic behaviour pattern of the country's leaders. The Party (UNIP) and its Government was searching for ways of containing a new social order of self-reliance and egalitarianism. Its leaders were therefore expected to mirror the nation's good image and not to use their acquired positions to amass wealth and exploit others. In other words, any leader who adhered to the Leadership Code would apparently be not placed in any doubtful position from the point of view of himself and his followers. Leaders were supposed to declare their property and were prevented from having more than one source of income.

However, definition of and restrictions of those considered as leaders under the Leadership Code somehow subtracted from the assumed degree of shared responsibility for development between the state and the country's private individuals. Zambian individuals likely to be affected by the Leadership Code were the very people with the right acumen to take the needed initiative. Their exclusion therefore could only have left the sharing of development responsibility to the state, the parastatals and non-Zambians..

FOREIGN INVESTMENT

Developing countries the world over suffer from the lack of means for capital accumulation. This is a continuous problem in Zambia, especially given the country's dependence on the export earnings from primary raw materials. These have to be supplemented by external aid to sustain many investment projects. There is no denying that in countries like Zambia foreign investment and assistance can play an important catalyst role in fostering the process of economic development. This is also true even for the developed countries. For instance, West Germany and Japan after the Second World War, were salvaged by foreign investment under the Marshall Plan.

The idea of indigenous participation is by no means as strange as many people outside Africa would like the host African countries want them to believe. There is therefore no mystery why foreign investors and their investment should appear to be alarmed with the trend. Many international companies have already come to terms with the notion and prefer teaming up with the host governments or nationals of the countries in which they invest. There is no country where the state does not exercise some control on foreign-owned companies.

Five arguments on the merits of foreign investment will be reviewed in this analysis. It is generally assumed that foreign investment contributes to a recipient country's development programme by helping reduce the shortage of domestic savings while increasing the supply of foreign exchange. The second argument is that foreign investment is a way of filling gaps in the domestically available supply of savings necessary to achieve development targets. Third, it fills gaps left by any shortfall in foreign exchange earnings from exports. Fourth, it provides for a transfer of technological skills, entrepreneurship, organizational experience, innovations in product and production techniques and managerial personnel. Fifth, protagonists hold the view that with time, foreign investment increases the resulting real income at a rate greater than the resultant increase in the income of the foreign investor.[15]

Answers to some of these questions are of great interest for developing countries eager to industrialize their economies. In their efforts, they encourage and attract direct foreign investment through offering a variety of incentives. This is why developing countries have worked out a number of incentives. The incentives are such benefits as tax concessions and subsidies, accelerated depreciation concessions, tariff protection, special facilities like offering or developing free industrial estates and providing public services at highly discounted rates.

The economic ventures undertaken in Zambia with extensive direct foreign investment include, among others, the Livingstone (Fiat) Motor Assembly, the Tika Iron and Steel Plant, the Kafue Fertilizer Plant, the Zambia Engineering and Construction Company (Zecco). The magnitude of benefits to the economy from direct foreign investment in the first decade of independence (excluding the mining sector), when measured in terms of the five criteria do not tally well. In other words, there is a lot of explaining required from the foreign investors as to concealed motives. Some of the so-called direct investments actually obstructed Zambia's development process during the period under review and beyond.

The analysis will test the validity of the above assumptions against Zambia's experience in the first decade of independence. The question is whether direct foreign investment was beneficial to the Zambian economy. Did foreign capital fill gaps between foreign exchange requirements and those derived from net export earnings and did it contribute to foreign exchange savings? Was there a proportion of foreign subsidiaries' products sold in the host country so that there would be less dependence on imports, resulting in greater foreign exchange savings?[16]

It will be illustrative of some of these facts to review the operations and impact of the Livingstone (Fiat) Motor Assembly project on the Zambian economy. A motor assembly plant of the type established in Zambia is normally a one-shot economic affair. Thus, it is an act with a limited multiplier effect, since the main employment components are imported. Consequently, there is very little that ensures commensurate benefits to the host country. On the contrary, it ensures a continuous and protected drain of foreign exchange from the recipient to the developed country. There are limited possibilities for increasing the number of indigenous employees at the plant, even the small maintenance squad cannot really be expanded. The price and value paid for imported skills was kept high. The additional services were overpriced and always paid for in foreign exchange.

The goal of conserving foreign exchange because the Fiat cars were assembled locally was not met. First, Zambia could not have saved foreign exchange because all claims for franchise and management fees to the parent Italian Fiat Company were settled in foreign currency. Second, the price of assembled components supplied from Italy was determined by the supplier and Zambia, as importer, had no way of knowing whether the prices quoted were for old or new stock. The extent to which Zambia benefited from the large size of the project is debatable. Fiat deliberately avoided manufacturing spare parts in Zambia as that would endanger the jobs of Italians in Milan. Fiat's behaviour in Zambia differed considerably from that in similar operations in the more industrially developing countries of Latin America or in Yugoslavia, where definite transfers of technology were insisted upon and received. Zambia found itself suddenly saddled with pressures created by return flows of interest, profits and dividends on accumulated investments and repatriation of capital thereby culminating in serious balance of payments problems. The local labour content was of far less economic significance in terms of its value added to the final product that desired.

It must therefore, in the light of the above, be discounted as a contributing factor to development or as a source of taxable income. In other words, this foreign investment became an effective way of making Italy richer and Zambia poorer.[17]

It was a paradox of the Fiat episode that the locally assembled car was relatively more expensive than its imported counterpart. It is the nature of multinational corporations like Fiat to keep the periphery economies in a state in which the latter cannot challenge the former's economic supremacy. This is not to deny the fact that an economy can derive benefits from foreign investment, especially when the investors are dealing with a host country in which the former is able to identify a long-term commitment.

Apart from the five economic ramifications of foreign investment, there is the aspect of social opportunity cost or the human or social welfare. Social welfare can be described as the well-being of the community as a whole. It is very difficult to measure. For one thing, it cannot just be measured on the basis of a simple equation such as income and output per capita as seemed to be implied.

The utility of an individual is difficult to measure and more so utility of the community. The issue ought to be examined from the point of view of certain measurable yardsticks. For the purpose of this particular analysis it could possibly be assessed against payment made to the Zambian local factors of production and the Government itself. Simpler formulas of social opportunity cost are easy to understand when taken together with the value-added approach of the kind given in this equation. The contribution (S_z) represents the multiplier effects arising from use of local inputs and job opportunities created. The resultant S_z is equal or higher than repatriation payments*

$$S_z = F_j (P_j + O_j)$$

where S_z = direct contribution to the economy
P_j = payment to factor (j)
O_j = social opportunity cost of (j)
z = Zambia

The usual temptation in dealing with mathematical equations and social behaviour, especially in the light of the foregoing, would be to ask whether economic ventures such as the Livingstone Fiat Motor Assembly, with a direct foreign investment, can in fact pass the acid test of having contributed commensurately to Zambia's economic development. Its impact was complicated by the lack of congruency between the parties. Whatever advantage could have resulted from foreign investment in aiding the growth

* For a discussion of this measure see the case of Liberia by Aimesh Goshal, "The Impact of Foreign Rubber Concessions on the Liberian Economy, 1966-71," *Journal of Modern African Studies*, 12, 4, 1974, pp. 589-599 (*i*) Y = X plus D plus F (*ii*) $V_z = Y - M - Z$ where Y = receipts, X = exports, D = domestic sales, F = foreign capital inflow, M = use of imported goods and services and Z = use of local goods and services.

of industrialization in Zambia was lost because of liabilities resulting from repatriation of enormous earnings in the form of profits and dividends and failure to transfer know-how and skills to indigenous Zambians beyond the few who worked in the factory.

It would appear on the surface, at least, from the above that a major principle of roads to development is harmonious development of all sectors of the economy. In other words, a need for effective monitoring and guidance in order that some sectors do not over-develop at the expense of others.

On the basis of the limited information available, and in the light of the rather limited period chosen for this analysis (i.e., from 1964 the year of Zambia's independence to first half of the 1970s), the overall economic performance of several Third World countries does not give much room for optimism. Such pessimism is compounded in the late 1970s and early 1980s by the dramatic increase in population and labour force, coupled with less increase in production. A combination which contributed to stagnation of the GDP in real terms. The extent to which planning was and could be an effective instrument for upholding the economy is perhaps the yardstick against which the economy is able to deal with the dramatic or destabilizing forces of price falls and instability; fall in the level of gross investment in constant prices; unemployment; income growth and welfare inequality. These are analysed at the end of the concluding chapter.

REFERENCES

1. "What Now: The 1975 Dag Hammarskjold Report on Development and International Co-operation", *Development Dialogue*, Fourth printing 1978, p. 7.
2. Chinese proverb.
3. John H. Adler, "What We Learned About Development?", *Finance and Development*, Vol. III, No. 3, September 1966, p. 159.
4. Gloria Nikol, *African Woman*, No. 9, March-April 1977, p. 9.
5. Samir Amin, "Growth is Not Development", *Development Forum*, Vol. 1, No. 3, April 1973, p. 1.
6. Irving S. Friedman, "Dilemmas of the Developing Countries—The Sword of Damocles", *Finance and Development*, Vol. 10, No. 1, March 1973, p. 14.
7. Peter E. Temu, *Marketing Board Pricing and Storage Policy*, East African Literature Bureau, 1977, pp. 117-120.
8. Republic of Zambia, *Second National Development Plan January 1972-*

December 1976, Ministry of Development Planning and National Guidance, December 1971, p. 6.
9. Joram Mayshar, "Should Government Subsidize Private Projects?" *The American Economic Review*, Vol. 67, No. 2, March 1977, pp. 20-28.
10. Nachum Finger, *The Impact of Government Subsidies on Industrial Managemer.i: The Israeli Experience*, New York, 1971.
11. Henry S. Meebelo, *Main Currents of Zambian Humanist Thought*, Oxford University Press, Lusaka, 1973, p. 4.
12. M. Lungu, "Co-operative Efficiency in Zambia", *African Co-operatives and Efficiency*, Carl Gosta Widstrand (ed.), The Scandinavian Institute of African Studies, Uppsala, 1972, pp. 207-226.
13. Father Bede Onuoha, *The Elements of African Socialism*, Andre Deutsch Ltd., London, 1965, p. 73.
14. *The Daily Mail*, Vol. 1, No. 2, 204, Lusaka, May 10,1977, p. 4.
15. G.M. Meir, *The International Economics of Development*, New York, 1968, p. 138.
16. T. Ishimine, "Performance, Motivation, and Effects on the Balance of Payments of Japan's Direct Investment", *Economia internazionale*, Vol. XX 1 (1-2), February-May 1973, p. 55.
17. P.A. Baran, P. Sweezy, "Notes on the Theory of Imperialism", *Economic Imperialism*, K. Boulding, T. Mukerjee (eds.) Michigan 1973. p. 164.

Conclusion

DEVELOPMENT: THEORY AND PRACTICE

The analysis has demonstrated the diversity of inherited theories of development and economic growth in Africa against the Zambian experience. It has also made clear that many of the countries linked the rate and direction of internal socio-economic change to export markets and with imports of skills, technology, capital goods and services and modern consumer products. For many a country, it had been hoped to use them to accelerate economic growth and diversification of the economy, reinforce existing patterns of production by manipulating exports and imports.

Complications arose as soon as policy-makers fell into the trap of mistaking modernization for development as if the two were easily interchangeable. Over and above, the inability to cope with the pluralistic structure in which they had found themselves at independence. That is, a traditional subsistence sector on one hand, and on the other, an apparent indigenous monetized sector controlled by a strong foreign enclave.

The ability to achieve development goals is what the Economic Commission for Africa assumed to be the first major crisis in socio-economic policy-making for most African countries at independence. This has also been reflected in the difficulty encountered in implementing their respective development strategies. Elements of the crisis include the relatively low elasticity of demand for agricultural export products; the temptation to expand production and exports in an effort to compensate for falling prices; the differing elasticities in the production of crops and annual crops' response to changes in export demand; the emergence of substitutes; an underdeveloped industrial base; and the inability to redress the urban-rural imbalance.[1]

The patterns of economic development chosen by developing countries in the early post-independence era have varied. This has been briefly alluded to in the preceding chapters. Two of these models stand out prominently: The market system (the capitalist

Conclusion

system which allows for market forces of supply and demand) has at times been condemned out of hand. The antithesis to the above, that of a planned economy (scientific socialism) has been applied among the developing countries who have shown a more socialist leaning. But not even among this group has there been a total rejection of the former.

Meanwhile, the high priests of the socialist-line of thought have had time to reflect on the economic plight of the main socialist camp—the USSR and its satellites, such as Poland. In particular, that these countries have proved not only inefficient but also incapable of feeding themselves. Often they have only been saved from starvation by imports of staple cereals and animal feed from the capitalist countries of the West.

The bottom line in Zambia showed beyond any doubt that it was a greater mix of the capitalist way of organizing production which enabled the country's economy to avoid sinking into serious economic trouble. Centrally-planned economies, based on the theories of Karl Marx are being used in several Third World African countries. The degree of success in those countries is very limited. Many of the latter countries have already begun to turn to the former. But it does not appear to have much appeal among Zambians.

Perhaps the Third World countries should be cautious about the use of socialist rhetoric, especially when they run their economies as inefficient and underdeveloped capitalist outfits. They create problems by disillusioning their people, whose crippling poverty increases daily, despite egalitarian slogans. Attempts were made in Zambia, under the FNDP and SNDP, to introduce new and relevant elements in economic planning. Some of those elements were perforce, a direct reaction to the effects of the UDI. The emergence of public corporations such as Findeco, Mindeco and Zimco and strengthening of Indeco were part of this process.

The Zambian experience is not atypical of difficulties faced by developing countries in integrating their economies. It was nevertheless fairly representative of what was stressed by King Moshoeshoe II about fulfilling economic aspirations. Thus he observed that it was not enough for Africa to have acquired flags and national anthems at independence, nor for African countries to flex their political muscles in international circles. It is much more important for African countries to actively and realistically work out appropriate means by which to stop the economic exploitation that they suffered and which was not fully redressed by the removal of the foreign political yoke.[2]

Zambia's future prospects is a sum total of more variables than have been analysed in this book, The future will depend on the interaction of all these variables. We have analysed some of the options available for economic growth and development in a formerly colonised country. Using the analysis, we can draw up certain presuppositions on the country's future.

Development is an intricate process. It entails considerable improvement in the material and social well-being of the people. The analysis has shown some of the dangers which arose during Zambia's first decade of independence. Misconceptions include equating development with non-development activities. For example, confusing development with the trappings of modernity such as building multi-storey office blocks in the capitals of respective countries, which in the final analysis only benefited a very small percentage of the population. These cannot be the central issues that concern economic development.

Central issues concerning economic development in most developing countries include questions of income redistribution, making the economy self-reliant and self-sustaining etc. In recent times, a fair number of economists are incorporating the concept of income distribution in the early stages of a proper development process. They have observed from various analyses that there is a tendency to confuse economic growth with economic development. Economic growth and development have tended, in many developing countries, especially in the early stages, to be at the expense of the lowest income groups. This trend was one of Robert McNamara's key points stressed in Chile in 1972 during the Third Session of the United Nations Conference on Trade and Development (UNCTAD).

Collaborative evidence is provided by scholars like I. Adelman and C.T. Morris. They selected some 44 countries in estimating percentage shares of income distribution in total national income going to population groups of different income levels. The analysis revealed that countries which had registered significant growth rate gains in GNP (Brazil, Mexico and India) were plagued with severely skewed income distribution patterns. In other words, more growth had resulted in declining shares of national income

Conclusion

going to the poorest 60 per cent of the people.[3]

Equitable income distribution is absolutely imperative if the development process is to proceed in any meaningful manner. Policies whose effect is to favour the rich at the expense of the poor are not only manifestly unjust, but in the end are economically self-defeating. They push frustrations to the point of violence, and turn economic advance into a costly collapse of social stability.

Zambian attempts to address this issue included direct government action in the form of promulgating of fiscal, price and income and rural development policies as well as by designed sectoral and provincial equalization. Direct government participation in investment were important, especially in determining the share of national income that must go to the various income groups.

Protestations against the capitalist models of pre-independence times and a bias towards a middle-of-the-road approach in post-independence era led to the decisions taken by Zambian policymakers. However, having been imbued in Western traditions led to use of some western development paths. The economy had been conditioned into producing for export what was not locally consumed and consuming imported rather that locally produced goods and services.

The Zambian economy was thus less able to become self-reliant except with certain drastic restructuring. The Zambian economy continued with its reliance on the inflow of foreign capital and skilled manpower. The early cushion made possibly by favourable terms of trade from copper revenues seemed to have blinded the nation from initiating viable diversification measures, as shown by the neglect of the country's agriculture sector.

In a nutshell, the country's economic development plans used a mixture of several borrowed development models. Elements of the inherited traditional Western system of economic growth and development persisted. Hence, the tendency to link the rate and direction of internal socio-economic change to export markets and the import of skills, technology and capital.

We could now focus attention on some of the needed structural changes. In particular, the country's future parastatals and other key economic elements. It is not possible to prescribe remedies for all the problems. However, the problem is great enough to demand bold action.

PUBLIC ENTERPRISES AND FUTURE DEVELOPMENT

State enterprises are an area of crucial importance to developing countries' future rate of economic growth and development because they are capable of generating or retarding national initiatives. They create or find themselves, by virtue of their legal economic role, with several forward and backward as well as vertical and horizontal linkages with other sectors of the economy. However, the application of the concept of state trading organizations (STOs) varies from country to country. The image of the Zambian public enterprises and their counterparts in Tanzania in the period under review, notwithstanding the success of some of their parts, left much to be desired.*

The role of state trading organizations in the economies of many developing countries is poised to grow in the future. This is bound to be the case for those countries which have shown a tendency to play down capitalistic models of fashioning economic development. Parastatals have experienced a rapid growth in the volume of their transactions, an increased amount of economic activity, in the value of their accumulated assets and the extent to which they were able to absorb unemployed labour.

The Zambian public enterprise like its Tanzanian counterpart, was basically state monopoly. There was no evidence of direct involvement of the workers in any of them nor was room made for such involvement to take place. They were public enterprises only because of direct treasury funding. The views of the general public were rarely listened to nor was there provision for accountability to the latter. Perhaps some mechanism should have been provided through which the general public could have criticised these institutions. As it was they tended to remain the monopoly citadels of the respective managers who were directly answerable to government machinery, especially the cabinet.

Needed structural changes should aim at creating greater possibilities for encouraging and increasing mobilization of national savings. The other side of the coin would be to encourage direct and indirect reduction of the burden on the state. In other

---

* For instance, the take-over of the sisal industry in 1967 and vesting its operations as a State Corporation has proved with the passage of time, to be rather uneconomical. Similarly many of Zambia's Findeco subsidiaries.

words, structural changes would force these institutions to generate their own working resources. A kind of new structural ownership, including equity participation of the indigenous people, would inevitably expose these enterprises to greater discipline, by virtue of market forces of supply and demand. Accountability to shareholders would inevitably bring these institutions face to face with the realities that they can survive only if their operations attain a high degree of efficiency. Under such circumstances they would no longer provide shelter to individuals who had failed in the political arena of the party and its government, which was so often the case.

The above proposed changes imply less reliance on the state. It follows therefore that it will require redesigning some of the old procedures. For instance, it would no longer be necessary to rely overwhelmingly on the "public purse". In other words, subvention by the Ministry of Finance whose tax or revenue collection base is the taxpayer or through external borrowing. This is particularly important given Zambia's rather limited tax base.

It is therefore in the national interest that the state in future, share with the public both the ownership and running of parastatals. The state can take precautions to avoid individuals becoming disproportionately more wealthy than others. The state must be content to remain a major minority shareholder in most of these public enterprises. Consequently, serious consideration should be given to encouraging each economic section, agriculture, banking, mining etc., to be directly involved and have a voice in their respective public enterprises. For instance, farmers and other indigenous Zambians should be invited to buy shares and invest in the Nitrogen Chemicals of Zambia Limited (the Kafue Fertilizer Enterprise) because its product is one of their very important needs. The people living near or growing cane sugar around the Nakambala Sugar Scheme should be encouraged to become shareholders and establish small holdings. Farmers should also become shareholders in the National Agricultural Marketing Board (Namboard) which is an agricultural nerve centre by virtue of its being a buying and distribution channel for farm products. A similar approach could be adopted in an attempt to correct some of the weaknesses observed with regard to the non-viability of the policy on cooperative development under both the FNDP and SNDP.

bours and eventually to the rest of Africa. There was still a lot left undone with regard to the country's telecommunications network by the close of the first decade of independence, especially considering its importance as a pillar for sustaining economic development.

Economic planners now consider the availability of electricity on a large scale as a logical departure-point in evaluating development. Demographers are in the nature of quoting the electricity component as part of the phenomenon of population movement of most dual economies. It is not uncommon for demographers and town planners to refer to the attraction of bright lights in urban areas as an explanation for rural depopulation trends.

To help reduce rural depopulation, as well as provide for economic development, electricity was introduced to 13 different rural centres. Making electricity available as well as other sources of energy, was an indispensable part of the country's industrialization process. The availability of energy in sufficient quantities greatly influences a country's rate of economic growth and development. The dilemmas faced by the country in this field were not only the enormous capital required but also reconciling major ecological factors. This included moving settled families and trying not to destroy the existing political goodwill. The story surrounding the construction of the Kariba Dam and the force used to move the people from the area, especially the resistance encountered by the administration of time, is a case in point.

THE ISSUE OF LAND IN DEVELOPMENT

Classical economists view the element of land as a necessary production component that must go hand-in-hand with capital and labour. The definition and appreciation of each of these elements has been modified over time, especially with advances in technology. None of these factors are any longer totally inelastic. Similarly, their respective role in production is subject to several interpretations.

Arable land can be increased by reclamation of seas and deserts. Land is also made more adaptable for other uses by scientific innovation. The application of fertilizer can make non-arable lands productive in the same way, today's construction methods

These are some of the ways which could be tried in order to begin to mobilize domestic savings and obtain a greater involvement and participation of the rural sector. What other better way of demonstrating in a practical manner the concept of productive and management participatory democracy than this? The same pattern of market transformation could be advocated in the other sectors. These experiments have been tried with tremendous success in other developing countries. For instance, the Gujarat Fertilizer Company in India has proved viable and profitable after implementing these changes. Similarly, in South Korea, where the State voluntarily handed over by public sale to nationals, the viable and profit-making national flag carrying airline (Korean Airline). Legislation in both countries was structured so as to allow the government necessary leverage to monitor and ensure that these enterprises operated in the interest of the nation as a whole.

An arrangement of the kind suggested above would go a long way in removing certain fundamental weaknesses in Zambian public enterprises, particularly where the problems emanate from cumbersome political-administrative controls. Managers should be consumers of the product(s) and directly affected by the enterprises' failures and/or losses. The bottom line for many of these managers is their monthly contract and comforts derived from the trappings of office. Many of them do not equate their salaries with a full day's work. A well defined framework of suitable checks and balances could be legally instituted in order to prevent the danger of concentrating too great an economic power in a few individuals or families.

An effective role for public enterprises within the context of development will only be possible after resolving supervision and management problems. It has become quite clear that there is a need to take corrective measures in this area. Serious consideration ought to be given to creating a two- or three-tier board system to run future public enterprises. To a large extent, this would allow for cross-fertilization of ideas, disciplines and interests, especially where the general public acquired equity participation. The future of state trading organizations in Zambia and their contribution to economic growth and development is assured. However, their objectives should be made clear as to what they ought to offer to the nation as a whole.

Chart II illustrates the envisaged linkages of the proposed

SUGGESTED PARASTATAL ORGANOGRAM

```
                            CABINET
                               |
                       Supervisory Board
                           (ZIMCO)
                               |
   ┌───────────┬───────────────┼───────────────┬───────────────┐
 Mining     Agriculture   Energy/Transport  Finance      Commerce/Industry
Management  Management      Management    Management       Management
  Board       Board           Board          Board            Board
 (ZCCM)                                                      (INDECO)
```

Grassroot Initiative (Mining): RCM, NCCM, MEMACO, MINDECO

Grassroot Initiative (Energy/Transport): Energy Corp., NTC, Zesco, Indeni, UBZ, ZA, ZR

Grassroot Initiative (Commerce/Industry): NIEC, CCC, NCZ, Super markets, KGP

Grassroot Initiative: CSB, DPB, RDC, Namboard, Horticulture

Grassroot Initiative: ZSIC, FINDECO, DBZ, ZNCB

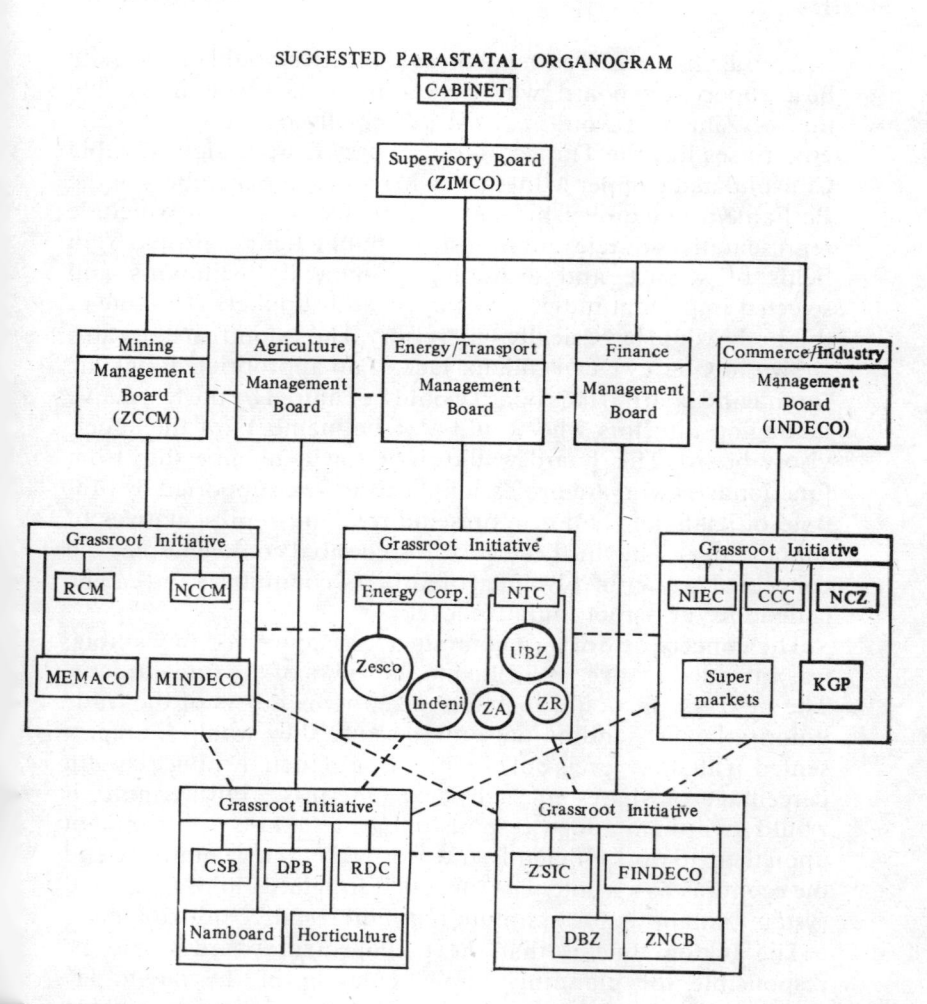

ZA   : Zambia Airways
Zesco : Zambia Electricity Supply Corporation
ZR   : Zambia Railways
ZSIC : Zambia State Insurance Corporation

KGP   : Kapiri Glass Products
NCZ   : Nitrogen Chemicals of Zambia
CSB   : Cold Storage Board
ZNBS : Zambia National Building Society
DBZ   : Dairy Produce Board

structural changes. The first of these categories would of necessity be a "supervisory board" whose coordinative task would resemble that of Zimco. It would consist principally of the head of the enterprises like the Transport and Energy Corporation. Zambia Consolidated Copper Mines (ZCCM) and some members of the Parliament with proven business experience. It would also include representatives of relevant ministries, public figures, people from fields of science and technology, financial institutions and selected important indigenous buyers and suppliers. The role of this body would be basically supervisory. The second category also shown in Chart we can call, for lack of an appropriate terms the "managing board". This board would be chaired by the respective managing directors who would also be members of the supervisory board. This board would draw for its membership from functionaries who are professional executives, supported by full-time officials and union representatives and representatives of shareholders. The third category, if so desired could embrace the concept of greater involvement of workers' committees and can be called the "grassroot initiative level."

The concept of workers' committees, as conceived in Zambia, was intended to give an avenue to the views of the trade unions. The new system would therefore give the members of the trade union a better standing not only because they would be represented at all three levels but also by virtue of their owning a certain percentage of shares in each such enterprise. Furthermore, it would give them added responsibility by making it incumbent upon them to work efficiently and profitably for the enterprise and the economy as a whole. The suggested structure builds on the old system but allows for optimum national resource utilization.

The bottom line is that the management board would be responsible for planning, implementation of the day-to-day activities of the public enterprise, while remaining responsible to the supervisory board which would have the prerogative of appraising and approving the former's activities. Enlightened legislation can take care of all of this. The Companies Act, apart from some of its clauses, is still basically rooted in the country's colonial past is insufficient to enact the changes. What is required is a change in the statutes covering parastatals to accommodate the new dynamism.

The future of such a system will have to rely on the power of legislation. Consequently, revamping of the legal structure will have to balance many interests, including aspects of an "economic payoff" to the nation as a whole. In this connection, the suggestion, as described in the Chart allows the government a supervisory and management role of all sectors of the economy. Thus, as in the framework of Zimco, the government would have ultimate responsibility. The government would also act as an active commercial participant and as an absentee owner, through its parastatals, could effectively influence the course of events on the management boards. Additionally, the creation of grassroot initiatives (that is situating correcting institutions like Chilanga Cement Company (CCC), Kapiri Glass Products (KGP), Menaco, Zambia Airways (ZA), etc. would ensure the participation of the private sector as well as identify with it in the short and long run.

REHABILITATION

Development economists are not surprised that the dilemmas faced at the outset of the attainment of political independence were not significantly met by the end of the first decade. In a nutshell, the country was still trying to change its monoculture economy. Copper, and the mining industry in general, continued to provide a disproportionately large share of the country's economic activities and its revenues. Other important sectors of the economy, such as agriculture and industry, continued to be relatively less developed. Income policies still tended to be directed at wages and salaries. As a result, the policies pursued were rather counter-distributionary by discriminating in favour of certain sectors of the economy. Industrial policy, apart from large investment in certain infrastructure, showed a greater bias for non-durable consumer goods and the processing of raw materials.

The fiscal policy and its adjustment had not narrowed the sectoral disparities existing between the urban and rural areas. The situation was made worse by the fact that the incomes generated by investments mostly went to or were retained by the urban sector. The high income earners did not set up projects in the rural areas.

The country's agriculture policy did not succeed in increasing levels of production. There were periods of food deficits in a country with much arable land, a favourable climate and several perennial streams and rivers. Some of the main stumbling blocks in this area included indecision over pricing policy. Attempts to establish state-run farms to produce food crops and dairy products were unsuccessful, as were attempts to produce industrial raw materials to sustain state enterprises.

There were successful attempts at producing wheat under irrigation but they were not sufficiently strong in view of the country's 2.8 per cent population growth rate and its rather high rate of wheat consumption. The 140,000-tonne capacity of the Kafue ammonium fertilizer plant (Nitrogen Chemicals of Zambia) turned out to be far below the country's 1985 projected 250,000 tonnes demand for fertilizer. Development of agriculture was also constrained by the inadequacy of storage facilities and marketing bottlenecks.

In economic terms, the rapid drift of rural dwellers to towns during the decade was caused by a familiar problem, the lack of an internal balance in the economy with respect to full employment, wage distribution and price stability.[4]

By the close of the country's first decade, the state had acquired considerable leverage which it could use in setting the pace of economic development. It had acquired the ability to raise and accumulate funds and it could have resorted to deficit financing in order to achieve some of its objectives. By 1974 the state had become a major proprietor in the mining, industry, trading, commercial and the banking sectors. Zimco was at one time rated the 54th largest company in the world. Surpluses in the sectors remained in the hands of the state for use in the development of other sectors of the economy.

The term "industry", for purposes of this analysis, covers Zambian activities pertaining to areas of mining and quarrying, electricity, manufacturing and construction. We have seen that industrial development is the key to rapid economic growth and to the transformation of the economies of the Third World countries, just as it was for the developed countries. The Zambian experience, in acquiring technology and developing a dependency on it, especially during the initial stages of industrialization revealed a number of things. In particular, dependence on technology led the modern sectors like mining and manufacturing to develop stronger linkages with the world economy, not Zambia's internal one.

Conclusion

Planners of Zambia's industrialization process should have taken account of experiences prior to and during the first decade of independence both within Africa and the Third World in general. Guidelines are provided in "A Programme for Industrial Development Decade for Africa" drawn up jointly by the United Nations Economic Commission for Africa, the Organization of African Unity and the United Nations Industrial Development Organization. In brief, it postulates that effective development requires necessary institutional infrastructures *inter alia*: the formulation of monetary and industrial policies, plans and programmes; project identification, preparation and evaluation; development or upgrading of traditional technologies; appraisal, selection, acquisition and adaptation of foreign technologies; regulation of technology, industrial financing, industrial consultancy, manpower and other services; standardization, testing and quality control; engineering and process design, industrial information; trading, etc.[5]

The empirical findings of this analysis confirm that developing countries should formulate development policies which stress self-reliance and self-sustaining economic growth. A country should identify and choose between the various options with a view to satisfying local needs, notwithstanding the fact that it is most difficult to bring about vertical and horizontal integration of Third World economies. Concentrating on processing and utilization, by the local industries, crops grown by the majority of the population in the rural areas is one of the solutions to this problem. It is also part of an import substitution strategy.

Development of agro-industries is also part of the import substitution strategy of industrialization. In developing, unlike the developed economies, the process has resulted in the creation of an industrial structure that became increasingly dependent on inputs of capital goods (principally imported machinery), spare parts, intermediate products (a greater part of which could, in the long run, be locally substituted), expertise and other supplies. Obviously, some of these import substitution initiatives should again be brought to the fore. Examples of projects with the above problems are the Nega Nega Brick, the Tika Iron and Steel Project etc. However, the most classic agro-industry failure was Indeco's decision in 1973, to close the jute milling at the Kabwe Fabrics Factory. Primarily because of the persistent folly of parastatals among developing countries failing to build up their own pool of indigenous consultant firms that could be used to find solutions to local problems.

Events surrounding Kabwe Industrial Fabrics assume greater significance because their adverse effects span the first three post-independence development plans. My research reveals that Indeco took the decision to halt the production of the jute mill on the basis of an 11-week (foreign) experts' report. The report in question, by Agro-business of New York, stated that locally grown kenaf was expensive compared to what the Kabwe Industrial Fabrics could import. However, there is no mention in the experts' report recommending closing the mill.

It is important to know that the idea of setting up this project was mooted in the 1960s by the Government through the then Ministry of Rural Development. It was part of a kenaf development programme intended to supply raw material requirements of the Kabwe factory. The project was constructed at the cost of K 3 million. Machinery at the plant included 58 looms purchased from Mackie and Sons Company of Northern Ireland. They were later sold at a give-away price to Laem Thong Industry of Bangkok, Thailand. Positive arguments included the now discredited argument that every import substitution industry saves a country enormous foreign exchange. The kenaf project was, if implemented, to save Zambia about K 900,000 annually in foreign exchange by not importing bags but using locally produced bags from hessian and kenaf. It was also argued that it would expand the country's narrow spectrum of cash crops (maize, cotton, groundnuts, etc.) The closure of the jute milling sector of the Kabwe Industrial Fabrics somehow put the clock back because of its great bearing on the country's initiatives at linking the growth of the agricultural sector with the establishment of agro-industries.

In retrospect, rehabilitation of an economy or the development crisis, or what we have called the Third World countries development options, is illustrated by the challenges and dilemmas faced by Zambia in its first decade.

The government and certain indigenous individuals attempted to usher forth a new economic order by deliberately engineering linkages and interactions among several productive areas, such as those in industry, agriculture, commerce and the rural sectors. They, at the same time, allowed for a continuous flow of innovations through education, as a dynamic instrument of change. The dilemmas faced in the first decade are likely to repeat themselves unless past mistakes were avoided in a totally new framework. Changes to the pattern of ownership and indigenization are needed and certain structural features of the economy, especially those characterised by outward orientation, lack of skills to undertake feasibility studies and poor internal linkages, must be corrected.

The country may continue to produce for the external markets and reduce consumption of what cannot be produced locally. Reliance on a few export products, over which the country had little influence on prices paid on the world market has to be eliminated. Transit routes must be efficiently maintained to make export prices more competitive. Reliance on external assistance has to be moderated.

It is obvious that external assistance of any form cannot be a country's backbone in the development process. There is conclusive evidence throughout human history that true development must be endogenous both in nature and character. On the basis of the analysis above, Zambia seems thus far to have only paid lip service to adherence to the principle of endogenous development. The results of the first decade made Zambians more aware that development of their economy and all its elements remains essentially the task and responsibility of Zambians themselves. In other words, that destiny cannot be transferred to the so-called fairly gods of whatever piety of intention.[6]

Zambia's dilemmas of economic growth and development have similarities to those experienced by other developing countries facing persistent sectoral imbalances. For example, almost all industrial activity was concentrated in a few urban areas in Zambia (10 towns), in Kenya (four towns), and Tanzania (five towns). The difference between the lowest and highest cash incomes in Zambia in industry and agriculture was in the ratio of 1:12 and 1:50 respectively, and parallels that of Kenya (1:16 and 1:50) and Tanzania (1:13 and 1:50).

PERISCOPIC PERSPECTIVE*

It will by now have become obvious that the objective of deve-

* In further developing this end part of this conclusion, I owe my gratitude to Dr. Manenga Ndulo and his academic colleagues of the Department of Economics of the University of Zambia for their invariable constructive criticism.

lopment strategies is to make it feasible for countries of the Third World to initiate or accelerate internally located and relatively autonomous process of growth, diversification and integration. To this end therefore, realization of the objectives and targets of a development option at national level depends, to a large extent, on the many facts alluded to in the preceding chapters. In particular, on the supply and quality of entrepreneurship and management conceived in the broadest sense both within the public and private sectors. The availability and access to external resources in an interdependent world could continue to be of great importance in augmenting domestic resources but need not be allowed to assume a predominant role. Rather, the vision and competence of governments and communities should be sharpened to select particular technologies, types of skill, capital goods and services and combine them with local resources in order to determine the results to be achieved.

There is a time difference of almost a decade between the publication of this material (1985) and the completion of the foregoing analysis (1978). Since then, quite a lot has happened to the economies of most of the Third World countries. For many of them, economic characteristics had only undergone very insignificant changes. However, issues have arisen in the intervening period to either confirm or contradict certain orthodox and theoretical economic predictions analysed, especially in the preceding analysis.

We should perhaps briefly return to the analysis in Chapters Three and Seven. The concept and ideas about planning have considerably changed in recent years. We therefore must exercise some caution about prescribing possible prescriptions of the course to be followed by small economies. The biggest challenge is perhaps how they can effectively and coherently integrate what others call action-taking, knowledge-creation with institution building. Chapter Three contains references to many planning paradigms. For instance, the (blueprint) model which means devising a design for the future that is carried out by a central authority according to a specific programme. The social approach of planning on the other hand, according to John Friedmann and George Abonyi envisages dovetailing practice with theory within a single movement involving the following four dimensions (*a*) theory of reality; (*b*) political theory; (*c*) social values; and (*d*)

Conclusion

social action.[7] The question is where did small economies of the Third World countries go wrong after the glorious post-independence years?

One of the answers lies in the fact that future development prospects for the Third World countries like Zambia had become inseparable with the continuing recession in the developed industrialized countries. The degree of pessimism was heightened by the threat of protectionist measures. The end result for many was increased adverse terms of trade and a fall in export volume; high and more volatile interest rates that limited their borrowing capacity while their debt repayment burden continued to increase. Zambia's average annual growth rate of consumption and investment between the periods of 1960-70 and 1970-80 for the public and private sectors fell from 11 to 1.8 per cent and 6.8 to 2.2 per cent respectively. Similarly, the national gross domestic investment fell from 10.6 per cent to 5.6 per cent in the same period.[8] Hence, Zambia's external debt rose from US $ 581 million in 1970 to US $ 1,815 million in 1980. In Ghana, it increased from US $ 489 million to US $ 1,011 million respectively.

The time lapse, i.e., in the aftermath of the first decade of the independence honeymoon is instructive of the challenges to the Zambian economy. Colleagues at the Faculty of Economics at the University of Zambia among others, have pointed out that (most challenging problems facing Zambian policy-makers arose in the mid-seventies). This was not surprising. The epoch was a reflection not only of its own time but also of some of the foundations laid or taken in the previous eras.

Put the other way, the performance of the Zambian economy in the second half of the 1970s and the first half of the 1985, while not atypical, is similar to that of the other Third World countries. In particular, it is an example of a country beset by common economic malaise, especially its domestic problems. Similarly, the inability to increase exports and reduce imports, the ability to borrow to reduce the debt burden. Rather, the economy had to cut imports and face stagnation. In other words, performance was dependent partly on developments in international environment described in the preceding chapters. It was also partly dependent on domestic policies. Report after report, especially in the early 1980s by the various United Nations agencies and its specialized agencies like the World Bank[9] have attempted to analyse both the root causes and to put forward possible solutions to the problem.

According to the World Bank basic indicators, Zambia with a GNP per capita of US $ 500 in 1979 was among 17 other countries (Ghana, Kenya, Lesotho, Mauritania, Senegal, Angola, Liberia, Egypt, Cameroon, Zimbabwe, Congo, Morocco, Nigeria, Ivory Coast, Tunisia and Algeria) reckoned to fall into the Middle-income country bracket.[10] The group is by no means homogenous. It is a mixture of some semi-industrial and oil-exporting countries with those who are basically primary producers. These differences are also reflected in the differences of these countries' ways and capabilities to adjust.

Zambia was to all intents and purposes, at least in economic terms, a small economy but not a ministate. It could not be described as a ministate like say Lesotho or the Gambia because its population was many times greater than 1,000,000, i.e. over 6,000,000 by 1984. A simple definition of a small economy can be based on the characteristics of degree of openness to trade and capital movement, export dependence and because a country is a price-taker for its exports.

Price-taking behaviour and that of trade openness is at the root of many of the challenges. It confirms in part, why the share of exports and imports in countries like Zambia between 1975 and 1982 in GNP was well above 50 per cent. On the other hand, the average annual growth rate of exports between 1970 and 1980 had declined to 1.2 per cent as against 2.2 per cent during 1960 and 1970. The average annual growth rate for imports also dropped from 11 per cent to 10 per cent respectively during the same periods. The country's terms of trade, taking 1975 as base year, worsened in 1980 to 83 as against 119 in 1960.

The Zambian experience shows that there are several important consequences which follow from a high degree of openness of the economy, especially in the trade sector. First, on the country's domestic price levels, these became a function of movements in price levels of imports. Their impact extended to prices of even non-traded goods and services and was brought to bear by foreign prices on other cost movements (including wages). A second transmission belt was provided through imports which tends to produce effects similar to those on domestic prices by foreign prices. Thirdly, the trend of increased protectionism in the principal export markets through a variety of trade barriers, tariffs

Conclusion

and exchange restrictions caused adverse effects on production and prices. This, to a large extent, led to underutilization of available or installed capacities and increased unemployment problems. Fourthly and not least of all, a combination of the above was instrumental in conditioning the Zambian economy into producing very few export products, most of whose prices were subject to fluctuations in world markets.

This in itself is another manifestation of the weakness of most small and low-income countries. In other words, prices of primary exports like copper in the aftermath of the mid-seventies, reflected the danger of concentrating in commodities for which demand expanded slowly.[11] Especially where this is coupled with the inability of countries already heavily dependent on one or two exports to vary their export output mix when relative prices changed. Zambia was caught in the paradox of the 1970s: the sharpest volume increase was in metals and minerals but this was also the area where prices fell most.

The challenges of the time to which we require solutions are many and complex. Nonetheless, an indicative list could be suggested on the basis of some of these challenges. However, we intend to approach these issues in the form of questions with a view to enabling policy-makers to discover why the economy was compelled to experience:

— Stagnant levels of GDP in real terms since the 1970s. Could the decline in per capita indices of GDP, consumption, savings, etc., have been due to Zambia's dramatic increase in population and the labour force?
— Difficulties in securing replacement investment which resulted in net decapitalization of assets. Was it only due to the dramatic fall in the level of gross investments in constant prices between 1970 and 1980 or were there some other factors?
— Drastically reduced volume of real exports and subsequently of imports and the import capacity in real terms. In other words, was it all due to the fall in copper prices and the rising debt service burden? What of aspects of over-invoicing of import bills and the shadow of transnational (TNCs) or multinational corporations (MNCs)?
— A higher than normal rapid rise in inflation. It rose at such a high speed and even surpassed other African countries with

relatively less natural resources endowments. Inflation in Zambia was running unofficially at about 20 per cent annually (during the early eighties) as against the official and World Bank estimates of 7.6 per cent (1960-1970) and 8.1 per cent (1970-1980).

So many difficulties in restructuring the mining sector in spite of the fact that since the Matero Reform Programme, the government was the majority shareholder. What Zambia needs is not only the problem of restructuring but also those of rehabilitation and putting the sector into a profitable framework.

That its agriculture sector had failed to respond effectively in the seventies. It had become a major cause of concern in the first half of the eighties. The attempted analysis of agriculture, stresses the fact that it is a key feature of development experience. History of the Japanese and Korean experience and most industrialized countries is sufficiently illustrative of the strong association between advancement in agriculture and overall economic growth. There is ample empirical evidence that a major cause for sluggish economic development in most middle- and low-income countries like Zambia, is slow agricultural progress. The average annual growth rate of agriculture in the period 1970 to 1980 for the following sample countries was: Zambia about 1.8 per cent; Ivory Coast about 3.4 per cent; Kenya about 5.4 per cent; Tanzania about 4.9 per cent; and Malawi about 4.1 per cent. The Zambian situation would have been more bearable had the average annual production growth rates of the other sectors like industry, manufacturing and services been higher than that of agriculture. On the contrary, these were considerably lower than in agriculture: about 0.1 per cent (industry); about 0.4 per cent (manufacturing) and 1.2 per cent (services). In Ivory Coast on the other hand, with an annual production growth rate in agriculture of about 3.4 per cent, the other sectors were 10.5 per cent (industry), 7.2 per cent (manufacturing) and 7.0 per cent (services) and compare favourably with Malawi (7.0 per cent; 6.7 per cent; and 9.1 per cent respectively).

Perhaps it is at this point in the conclusion that we should reiterate the interlinkage existing between the various historical epochs. In particular, that the past (yesterday), today (persent) and

Conclusion

tomorrow (future) are a linear chain of events. In other words, Yesterday is the seedbed of Today and Tomorrow. Today is inescapably a reflection of the past (Yesterday) intertwined with current developments. Tomorrow is a sum total of events in the past (Yesterday) and those in the present (Today).

If experience (a sum total of yesterday and today) is any guide to future policy, I am reluctant to prescribe a single remedy or strategy. However, the short list of challenges in the early eighties calls for specific remedies answerable to relatively small economies like Zambia. Hence:

— A country with an abundance of mineral resources has to give priority to establishing an export industry based on greater utilization of those resources. In Zambia, it is no longer a question of restructuring the mining sector but making it profitable. Attention ought therefore to be given to finding ways of mobilizing not only external assistance but also private investors. Both can play useful roles in providing project design and feasibility studies within suitable legal frameworks, with a view to enhancing the country's overall economic well-being;
— Zambia as a land-locked country ought to establish a range of home industries. In other words, import substitution for certain products can be economic and deserve a place in the country's industrial strategy, which due to high transport cost can enjoy natural protection;
— A development of indigenous industries and skills also ought to be a crucial part of the strategy;
— Policy-making and planning of industry must not overlook assistance to a multiplicity of small-scale industries. In particular, assisting them to prepare bankage projects;
— Better and fuller exploitation of the agriculture potential necessitates the adoption of food processing industries on a priority basis. It could provide a natural link with other economic sectors, like manufacturing. Similarly, developing suitable links with agriculture would be good for small-scale operations that need to meet both domestic consumption and export requirements;
— Programmes for export promotion, especially export-processing activities will need to ensure economical backward and forward linkages with the development of skills and technical

assistance. At the same time, priority will need to be given to ways of encouraging domestic industries to produce inputs for local industries;
— The very nature of small economies is that they are rarely able to disassociate or delink themselves from external technological assistance deriving from their trading partners. But this dependence has to be mutually beneficial.

The examples in the earlier chapters have highlighted areas of possible conflict of interest between the host country and the foreign investors. In particular because the latter rarely purchase local services, like financial consulting etc. This phenomenon is not unique to the Third World countries. It is also prevalent in the industrialized countries, except in being relatively small in scale. For example, Uniroyal closed its inner tube plant in Indianapolis not because it was not profitable but because the plant operated marginally. Such revelations cannot be the only reasons why small economy countries should shy away from attracting foreign investment partnership. What is needed is to guard against the dangers of possible increase in economic instability which can be brought about by say closure of plants, or denial to other sectors of needed economic linkages.

Admittedly, the selection of "development options" from the above overview of the critical areas is a major factor in determining future economic prospects for most Third World countries. There is already a considerable amount of literature in this area and this has given rise to an apparent conflict between the various champions. On one hand, the advocates of orthodox development economic theory. On the other hand, the school which advocates the dynamics of concentration and marginalization.

Fortunately, for the world at large and also the countries of the Third World, the apparent ideological or intellectual differences do not dispute the need to solve problems of development: income growth of individual sectors and territories; employment creation and prevention of unemployment; achieving equality and raising general welfare; efficient productivity and distribution.

One lesson from experience with trends in the economies of the Third World countries is that the adoption of a development option, in the absence of other ingredients does not result in development. It has to be part of a total package of structural changes. It also requires changes to the methodologies and styles of decision-making. In other words, decisions on and implementation of projects at the lower levels must not simply replicate those at the centre. Decision-making is a technique which has to be taken by people who have the capacity and the right to induce richness into the process.

Smallness in itself is not a factor for continued underdevelopment. There are many examples of small economies that have grown into enviable economies. India's cottage industries is one good example. Writers like E.F. Schumacher have extolled the virtues of economies of scale that derive from small economies of scale in his book "Small is Beautiful". The lesson is for such small economies like Zambia, to adopt appropriate development options with built-in mechanism for adjustment bearing in mind the advantages of large-scale economies.

Africa's heightened economic crisis in the eighties implies among other things, an improvement in agriculture management, especially food self-sufficiency alongside industrial development on the basis of local resources mixed, as desired, with imported inputs. Many of these can be better developed, if there is greater utilization of foreign technical assistance in product design, processing and production techniques as well as distribution. A combination of these activities could substantially create more employment opportunities and lessen the vulnerability that arises from trade openness of many a Third World country.

REFERENCES

1. ECA, "Revised Framework of Principles for the Implementation of the New International Economic Order in Africa", E/CN.14/ECO/90/Rev. 3, 25 June 1976, para. 8.
2. His Majesty King Moshoeshoe II of the Kingdom of Lesotho, "Opening Address to the Fourteenth Inter-Africa Public Administration Seminar", *The Fourteenth Inter-African Public Administration and Management Seminar: The Prospects of the Indigenization of the Private Sector of African Economies,* African Association for Public Administration and Management, Maseru, Lesotho, 1976, p. 10.
3. I. Adelman and C.T. Morris, "An Anatomy of Patterns of Income Distribution in Developing Countries", *Development Digest,* Vol. IX, No. 4, October 1971, pp. 24-37.

4. Patrick O. Ohadike, "Bottlenecks in the African Labour Situation in Zambia", *Journal of Administration Overseas*, Vol. XI, No. 4, October 1982, p. 268.
5. United Nations, *A Programme for the Industrial Development Decade for Africa* New York, 1982, ID/287, p. 46.
6. Alexander B. Chikwanda, "The Economy and Personal Glory", *Times Review*, Sunday Times of Zambia, April 11, 1982, p. 3.
7. John Friedmann, "Planning as Social Learning" *People-Centred Development*, David C. Korten and Rudi Klauss (ed.), Kumarian Press, Connecticut, 1984, pp. 189-194.
8. Table 4, Growth and consumption and investment, *World Bank Report 1981*, Washington D.C., August 1981, pp. 140-141.
9. See also *Accelerated Development in Sub-Saharan Africa: An Agenda for Action*, The World Bank, Washington D.C., 1981.
10. "Table 1, Basic indicators", *World Development Report 1982*, Oxford University Press, 1982, pp. 110-111.
11. "For Africa, ... aggregated output of goods and services failed to grow in 1982, and contrary to expectations, in 1983, the growth rate was minimal, estimated to be a mere 0.2 per cent over 1982. Thus, since 1980, output per head has been declining in constant terms by about 10 per cent per annum", *1982-1983 Biennial Report of the Executive Secretary*, United Nations Economic Commission for Africa, E/ECA/CM.10/11, pp. 2-3.
12. See Mariol I. Blejer and Mohsin S. Khan, "Private Investment in Developing Countries", *Finance and Development*, Vol. 21, No. 2, June 1984, pp. 26-29.

Appendices

Appendix A
Summary of Some Zambian Minerals: Uses* and Location

1. Asbestos: a common descriptive name of a group of naturally occurring fibrous minerals with a high proved strength to weight ratio and resistance to searing temperatures.

(*a*) *Uses:* (*i*) rockets and missiles; (*ii*) Almost all its uses are as a processed fibre based upon length. Thus, longest fibres are used in such things as textiles, clothing, different types of packings, woven brake linings, clutch facing, electrical insulation materials; another type is used in the manufacture of high-pressure cement pipes for transporting water; another type for low-pressure cement pipes, flat and corrugated cement sheets, wrappings and insulation, etc.

(*b*) *Location:* Zambia is devoid of basic igneous rocks in which deposits of commercial asbestos are known to occur. The very few known occurrences of asbestos-type mineral do not seem to justify development because they lack certain geological formation to justify necessary investment.

2. Clays: a product of mechanical and chemical breakdown of parent rocks which fall into at least six classes comprising of kaolin, ball clay, fire clay, bentonite, fuller earth and common clay.

(*a*) *Uses:* Kaolin has many industrial uses and many new uses are constantly being developed but it is in common use in paper, paint, rubber, plastics and ceramics.

(*b*) *Location:* Common clays and shales are indigenous to Zambia like in many other countries of the world. The

---

* Information relating to description and uses in respect of most minerals listed has been derived from the *Bureau of Mines Bulletin 677*, "Mineral Facts and Problems", issued by the United States Department of the Interior.

geological occurrence of deposits of clay over much of Zambia reveals the presence of white kaolin clay suitable in the production of white china and porcelain located in the Shiwa Ngandu area near Chinsali in the Northern Province. They are also known to contain ceramic clays which are suitable for mixing with feldspar and quartz for the manufacture of stoneware. Large quantities of brick clay suitable not only for manufacturing bricks but also for tiles have been discovered in many other parts of the country. Biggest deposits are those found south of Lusaka, in dambos and depressions near Kabwe, the Copperbelt and other parts of the country of similar geological formation.

The Nega-Nega deposits contain some fire-clay.

3. *Coal:* has certain common characteristics with wood and peat in terms of the element composition of carbon, hydrogen, oxygen, nitrogen, sulphur, etc. Consequently classified in an ascending order or rank on the basis of the amount of fixed carbon increases, and the amounts of inherent moisture and volatile matter decrease: lignite, sub-bituminous, bituminous, bituminous and anthractic.

(*a*) *Uses:* mostly for industrial purposes like power generation and for both coking and non-coking. Not all bituminous coals with coking properties are suitable for metallurgical purposes.
(*b*) *Location:* discovered in Zambia about the beginning of the 20th century. Known coal areas up to the time of writing occur in about two main areas. First in the extreme north-east area near the Malawi border in the Luangwa Valley. Secondly, in the Zambezi Valley around the Kandabwe and Maamba area, north of the Kariba Lake and not far from Livingstone. Possibilities of further finds in other parts of the country are quite promising.

4. *Cobalt:* a silvery grey metal identifiable by its atomic number 27, an atomic weight of 58.94 and melts at 1,495°C.

(*a*) *Uses:* principally in heat-and-corrosion-resistant materials, high-strength materials and permanent magnets. Used also in

Appendices

hard-facing alloys for wear and abrasion resistance, and tool and die steels. General use of cobalt is in respect of: electrical equipment and supplies; aircraft and surface engines and parts; machine tools; construction and mining; paints and related products; miscellaneous chemical products; and as an addetative to soils as·a nutritive agent to produce crops for animal food.
(b) *Location:* produced as a copper by-product in most Zambian copper mines.

5. *Copper:* classified broadly on the basis of method used to refine the metal:
 (i) electrolytic copper is refined by electrolytic deposition;
 (ii) fire-refined copper is refined by the use of pyro-metallurgical process;
 (iii) electronic cooper is copper deposited as a cathode from leach solution.

(a) *Uses:* largely in electrical equipment and supplies: electric motors, power generators, motor generators and sets, dynamotors, fans, blowers, industrial controls, and related apparatus which require copper for their best electrical performance. It also holds its own and is widely used in underground lines apart from dominating its usage in the small gage wire market.
 Aluminium is copper's biggest rival, particularly in most highvoltage overhead power transmission lines.
(b) *Location:* seven major producing mines in 1970 were all within the Copperbelt area. Other copper areas include Bwana Mkubwa, Mtuga, Alliers, Nampundwe, Hippo, Kalengwa, Kasempa, Lumwana and Kansanshi.

6. *Graphite:* a form of crystallised carbon which has a grey to black metallic luster. At least three natural types: (i) lump, (ii) amorphous and (iii) crystalline flake.

(a) *Uses:* largest uses are for carbon raising in the steel industry and for foundry facing.
(b) *Location:* geological investigations conducted have shown several localities of graphite prospects which consist of those discovered in the upper lower Luangwa Valley and near Lundazi.

7. *Gypsum:* a naturally occurring mineral with great agricultural and industrial significance.

(*a*) *Uses:* like sand, is one of the commonest building materials in use all over the world. Crude gypsum is marketed for use in cement, agriculture, fullers and wall-boards. In agriculture, it helps to improve the texture of certain types of soils.

(*b*) *Location:* extensive prospecting carried out for gypsum confirm substantial finds in the Lochivar area situated in the Kafue Flats and about 42 kilometers from Monze.

8. *Iron Ore:* is the primary source of iron, the metal most widely used by man. The colour is often used to identify the principal mineral content: (*i*) hematite is the principal mineral in red ore and specular ore; (*ii*) magnetite ore is black; (*iii*) brown ore usually consists of geothite or limonite and siderite.

(*a*) *Uses:* (*i*) mostly for iron and steel making. (*ii*) the manufacturing cement and heavy-media materials. (*iii*) smaller quantities of iron ore go into the manufacture of ferro-alloys, paint, high-density concrete aggregate, ferrites and as mineral additive to animal feed.

(*b*) *Location:* deposits in varying degrees of mineral content are found over a much wider area in several parts of the country than many other minerals so far discovered. Potential deposit reserves of, (*i*) less than 50 per cent iron ore content have been located near Chibote in the Luapula Province; (*ii*) with about 64 per cent iron ore content near Mwinilunga in the North-Western Province; (*iii*) with about 65 per cent and 5 per cent at Lufubu and Lubungu respectively. Discovered reserve deposits with considerable lower iron ore content potential were located in the (*iv*) Copperbelt area, ranges between 0.1 and 0.2 per cent of iron ore content; (*v*) The Sanja location near Lusaka with between 49 and 62 per cent; (*vi*) the Nambala site has between 40 and 60 per cent; (*vii*) while the Chisasa site is estimated between 35 and 60 per cent.

9. *Lead:* a soft but heavy nonferrous metal which is most corrosion-resistant of all common metals.

(*a*) *Uses:* fall under two categories: (*i*) metallic, where it can be

used alone or as an alloy with other elements and largely in the manufacture of storage batteries and vital in transportation, communication and electric utilities, (*ii*) chemically as a form of chemical compounds as an active component of gasoline anti-knock additive, and (*iii*) other uses include construction works, communications, industry in respect of trouble-free transmission, radiation shelding.
(*b*) *Location:* known deposits which continue to be exploited commercially are those at Mutwe-Wa-Nsofu near Kabwe.

10. Limestone (CaCO$_3$): a calcite which, when crushed and heated in specially constructed furnaces becomes quick lime (calcium oxide, CaO). Raw materials used for manufacturing lime are relatively cheap, however, the cost of processing is rather high. Transportation is one big factor in the cost of lime and determine the location of a lime kiln.

(*a*) *Uses:* to a large extent as a chemical reagent in (*i*) many industrial processes, (*ii*) in the building industry, and (*iii*) in agriculture.
(*b*) *Location:* plentiful and mostly along the Zambia railway line north and south of Lusaka.

11. Manganese: according to the Oxford Dictionary is a black mineral used in glass-making. However, no standard industrial specifications of it exists even by the authoritative United States Bureau of Mines. However, the term manganese ore is used for those ores containing a certain percentage of manganese.

(*a*) *Uses:* essential for the production of virtually all iron and steel and so far not substitutable. It can be alloyed with copper for the manufacture of manganese bronzes for ship propellers or other manganese-copper alloys, etc.
(*b*) *Location:* reserves with a mineral content of (*i*) about 50 per cent have been located at Kampumpa north of Kapiri Mposhi towards Mkushi in the Central Province; (*ii*) another mine had been worked to temporary exhaustion at Chiwefwe near New Mkushi Boma.

12. Mica: a group name for a number of complex, hydrous potassium silicate minerals with differing chemical composition and physical properties, selective product classifiable into several grades depending on size and quality.

(*a*) *Uses:* consumed mostly (*i*) in the fabrication of vacuum tube spacers, (*ii*) the manufacture of capacitors and (*iii*) washers which act as insulators in electronic apparatus. World demand is projected to decline due to technological changes and as a result of an increase in the development of substitutes.

(*b*) *Location;* small amounts of deposits from isolated prospects in the east-north-eastern part of the country are known to have been produced since 1920. A known by-product which has been produced in recent times has been beryl.

13. *Phosphate:* the earth's crust is said to contain about 0.23 per cent of phosphorus (P_2O_5) which happens to be the chemical analysis of phosphate rock of the apatite group.

(*a*) *Uses:* (*i*) apatite is calcium phosphate which is a major source of fertilizer. (*ii*) one typical occurrence of which is found in the carbonite deposits commonly contain the columbium bearing mineral known as pyrochlots and (*iii*) monazite which usually contains thorium and other minerals of potential economic interest as magnetite.

(*b*) *Location:* several known carbonatite deposits have been established. Investigations of their composition have not been exhaustively undertaken to determine all the possible varieties, particularly their respective ore content and amount of reserves. However, the Kaluwe deposit east of Lusaka was identified to contain enough mineable apatite.

14. *Sand and Gravel:* both are products of the weathering of rocks.

(*a*) *Uses:* (*i*) principal construction material, (*ii*) as aggregates for concrete and bituminous mixes and (*iii*) as fills for foundation stabilization. (*iv*) However, sands suitable for glass manufacture are characterised by very rigid specification of (*i*) grain size and (*ii*) chemical impurities.

(*b*) *Location:* geological tests of the Muva quartazite on the

Kapiri Mposhi Hill had established that with proper sizing and removal of mineral impurities (i.e. non-quartz grains) that they would meet industrial glass sand specifications. Since then the Kapiri Mposhi Glass Product plant has been established.

15. Tin: primary or virgin tin is metal produced from ore with its quality specifications determined principally by brand names that indicate the smelter and source of the ore.

(*a*) *Uses:* generally consumed in (*i*) metallic form and (*ii*) only to a limited extent used in industrial chemicals.
(*b*) *Location:* deposits of cassiterite concentrates have been produced in Zambia since 1935 from gravel deposits or deeply weathered pagmatitic veins and these have been located in (*i*) the Choma District and (*ii*) the area just a little south of the town of Mbala (formerly Abercorn).

16. Uranium: a naturally occurring silvery-white metal consisting of three distinct semi-stable radioactive isotopes.

(*a*) *Uses:* an important energy source because fission of isotope U_{235} releases large amounts of energy (*i*) Reactors for generating electricity have been developed into commercial power plants and light-water-cooled converter reactors as electric generators since 1960. (*ii*) Also required for fast breeder reactors apart from (*iii*) its military use in terms of nuclear war.
(*b*) *Location:* uranium mineralization has been recorded from a variety of rock formation in Zambia. Traces of both primary and secondary uranium minerals are recorded from pegmatitic belt of the Eastern Province. Furthermore, the Lower Roan biotite-talc-sercite schists and scapolite-kyanite beds in the North-Western Province have yielded significant uranium mineralization. However, the most important uranium province in Zambia is situated in the mid-Zambezi Valley in the rocks belonging to the Upper Karroo continental sediments. Uranium mining at Nkana's Mindolo mine is in association with copper mineralization.

17. Vanadium: grey to white metal, malleable, ductile metal usually produced as a by-product or co-product of another

element has been reckoned by geologists to be one of the more common trace elements in the earth's crust.

(*a*) *Uses:* (*i*) principally as an alloying element with steel in order to improve the strength of steel; (*ii*) as an alloy for heat-resistant or high-strength applications; and (*iii*) notably in sulphuric acid production and petroleum refining.
(*b*) *Location:* has been mined at Kabwe from the Broken Hill Mine since 1919.

18. Zinc: primary zinc is obtained from newly mined ore and is categorised either as electrolytic or distilled zinc, according to the reduction process used.

(*a*) *Uses:* four major areas (*i*) galvanizing, (*ii*) brass and bronze products, (*iii*) castings, and (*iv*) rolled zinc; (*v*) consumed also in pigments or other chemicals.
(*b*) *Location:* has been mined at Broken Hill Mines since 1919.

Appendix B
Summary of Zambia's Transit Routes
Brief Summary of Zambia's Main External Routes

Mode	Route	Description	Total Distance (km)
1	2	3	4
Railway	1	Tanzania-Zambia Railway (Tazara). Railway Head at New Kapiri Mposhi snakes northward via Nakonde to Dar es Salaam on the Indian Ocean	1,920
	2	(a) Zambia Railway from Livingstone via Lusaka, Kabwe, Kapiri Mposhi to Ndola, enters Zaire at Sakania to link up with the Benguela Railway to the Angolan port of Lobito on the Atlantic Ocean	2,902
		(b) Zambia Railway from the farthest Copperbelt towns runs south via Ndola, Kapiri Mposhi, Kabwe, Lusaka Monze, Livingstone to the middle of the Victoria Falls Bridge to link up with Zimbabwe Railways to the Mozambique port of Beira and South African ports on the Indian Ocean.	(2,379)
Road	3	From Livingstone via Kazungula to Nata in Botswana	393
	4	(a) From Chingola on the Copperbelt via Kapiri Mposhi and Nakonde to the Tanzania port of	2,038

1	2	3	4
		Dar-es-Salaam on the Indian Ocean (b) From Livingstone via Lusaka, Kabwe, Kapiri Mposhi and Nakonde to the Tanzanian port of Dar es Salaam	2,472
Road Railway	5	Lusaka as staging point by road, the route crosses into Zimbabwe at Chirundu to Salisbury to link up with Zimbabwe Railways via Umtali to the port of Beira on the Indian Ocean	1,040
	6	Lusaka as staging point by road, the route leaves Zambia via Chipata to Lilongwe in Malawi (1,107 km), then links to the railroad through Mozambique (for 560 km) to the port of Beira on the Indian Ocean	1,667
	7	Lusaka as staging point by road to Chipata (600 km) and on to Salima on Lake Malawi via Lilongwe (250 km), route links itself to a railroad through Mozambique to the port of Nacala (880 km)	1,730
Road/ inland waterway	8	From Kapiri Mposhi by road via Kasama and Mbala to the port of Mpulungu on Lake Tanganyka and by Liemba Lake Steamer to Kigoma on the shores of Tanzania to Dar es Salaam or Bujumbura in Burundi	
Air	9	(a)(i) Lusaka international airport, point for both international passengers and cargo traffic. (ii) Lusaka has a second airport	

Appendix B: 211

1	2	3	4
		of considerable commercial value whose capacity can be brought into immediate use if need be. (*b*) Zambia's second international airport with a built-in-commercial capacity is at Ndola. (*c*) The third international airport of potential commercial viability is at Livingstone.	
Telecommunications	10	Earth satellite station which provides Zambia with world-wide communications has been built at Mwembeshi outside Lusaka on the Mumbwa Road.	
Oil Pipeline		(*a*) Transportation of petrol up to January 1966 had been by railway (route 2(b) above) from Feruka Refinery near Umtali in Zimbabwe (*b*) The Oil pipeline connects Ndola to the Tanzanian port of Dar es Salaam started operating in 1968 with a capacity of 450,000 metric tons per annum using five pumping stations strung at intervals along its route	1,693

Bibliography

Brelsford, W.V., *The Tribes of Northern Rhodesia*, The Government Printer, Lusaka, 1956.
Bretton, Henry L., *Power and Politics in Africa*, Longmans Group Limited, London, 1973.
Cervenka, Zdenek, ed., *Landlocked Countries of Africa*, The Scandinavian Institute of African Studies, Uppsala, 1973.
Chileshe, Jonathan H., *The Challenge of Developing Intra-African Trade*, East African Literature Bureau, Nairobi, 1977.
--------, "Indigenization", *ENTERPRISE*, No. 2, Zambia Industrial and Mining Corporation Ltd., Lusaka, 1975, p. 43.
--------, "An Extension of the Development Paradigm: Grass Roots Experiments and People's Movements in Africa", *18th SID World Conference*, Rome, 1-4 July 1985
Elliot, Charles, ed., *Constraints on Zambia's Economic Development*, Oxford University Press, Nairobi, 1971.
Fortman, Bastiaan de Gaay, ed., *After Mulungushi: The Economics of Zambian Humanism*, East African Publishing House, Nairobi, 1968.
Franklin, H., *Unholy Wedlock*, Allen and Unwin, 1963.
Griffiths, Ieuan L., "Zambian Coal: An Example of Strategic Resources Development," *Geographical Review*, Vol. 58, No. 4 October 1968, pp. 538-551.
Harvey, Charles, *Economic Independence and Zambian Copper*, Praeger, New York, 1972.
Kaunda, K.D., *A Humanist in Africa:* Letters to Colin Morris, Longmans, 1966.
--------, "Towards Complete Independence", Address to the UNIP National Council, Matero Hall, Lusaka, 11 August 1969.
--------, "Zambia's Economic Revolution", Address at Mulungushi, on 19 April 1968, Zambia Information Services, Lusaka, 1968.
Makoni, Tonderai, "An Economic Appraisal of the Tan-Zam Railway", *African Review*, Vol. 2.
Makulu, H.K., *Educational Development and National Building in*

Independent Africa, S.C.M. Press, 1971.

Meebelo, Henry S., *Reaction to Colonialism: A Prelude to the Politics of Northern Zambia, 1893-1939,* Manchester University Press, 1971.

Mendelsohn, F., ed., *The Geology of the Northern Rhodesia Copperbelt,* Macdonald, London, 1961.

Mukupo, Titus B., *Kaunda's Guidelines,* TMB Publicity Enterprises Ltd., Lusaka, 1970.

Mwanakatwe, J., *The Growth of Education in Zambia since Independence,* Oxford University Press, 1968.

Northern Rhodesia Government, *Report on Intensive Rural Development in the Northern and Luapula Provinces of Northern Rhodesia 1957-1961,* The Government Printer, Lusaka, 1961.

Ostrander, Taylor F., "Zambia in the Aftermath of Rhodesian UDI: Logistical and Economic Problems", *African Forum,* Vol. II, No. 3, Winter 1967, pp. 50-65.

Ranger, T.O., *The Agricultural History of Zambia,* NECZAM, Lusaka, 1971.

Republic of Zambia, "Research Memorandum No. 11", Kabwe Regional Research Station, Department of Agriculture—Ministry of Rural Development, May 1970.

-------, "Annual Report of the Research Branch", Ministry of Rural Development—Department of Agriculture, 1970.

-------, "Annual Report of the Extension Branch", Ministry of Rural Development—Department of Agriculture, 1 October 1971—30 September 1972.

-------, "Nuclear Energy: Uranium Deposit of Africa". Occasional Paper No. 85, Zambia—IAEA Conference, Lusaka, No. 1, 1977.

-------, "Uranium Mineralisation in the Karroo System of Zambia: Uranium Deposits of Africa", *Occasional Paper No. 92,* Zambia—IAEA Conference, Lusaka, 1977.

-------, *Report of the Second National Convention on Rural Development, Incomes, Wages and Prices in Zambia:* Policy and Machinery, Kitwe, 12-16 December 1969, The Government Printer, Lusaka, 1970.

-------, *Village Productivity and Ward Development Committee.* The Government Printer, Lusaka, 1971.

Rousseau, Jean Jacques, *The Social Contract and Discourses,* J.M. Dent. London, 1913.

Bibliography

Rudden, Patricia, *The Story of Zambia*, Oxford University Press, Eastern Africa, Lusaka, 1968.
Sokoni, J. and Temple, M., *Kaunda of Zambia*, Nelson, 1964.
Stanford Research Institute, *Zambian Communications Study*, 1966.
The World Bank, *Accelerated Development in Sub-Saharan Africa: An Agenda for Action*, Washington D.C., 1981.
Turok, Ben, "Control in the Parastatal Sector of Zambia", *The Journal of Modern African Studies*, Vol. 19, No. 3, Sept. 1981, pp. 421-446.
United Nations, *Industrial Development in Africa*, ID/CONF-1/RBP/I, New York, 1967.
Welensky, R. *4,000 Days*, Collins, 1964.
Widstrand, Carl Gosta, ed., *Co-operatives and Rural Development in East Africa*, The Scandinavian Institute of African Studies, Uppsala—African Publishing Corporation, New York, 1970.
Zulu, J.B., *Zambian Humanism: Some Major Spiritual and Economic Challenges* Neczam, Lusaka, 1970.

Index

Abonyi, George, 190
Adleman, 1, 178
Africa, 81, 187; development plans in, 58; land-locked countries in, 41
African Government, 91
Africans, 80
Africanization *see* indigenization
Agriculture, steps to solve problems of, 121-22. See also agriculture under names of the countries.
Amethyst, 20
Angola, plight of blacks in, 80
Anglo American Corporation, 91, 110, 112; mining operation of, 118-19; recognition to, 91

Banks, 88, 101, 105, 108, 134, 135, 138, 168
Berlin Conference, 40
Bostock, Mark, 112
Botswana, 41, 155
British Foreign Office, 34
British South Africa (BSA) Company, 33, 34, 40, 91, 92, 110
British West Africa, 134
Burundi, 41

Canada, 147-48
Cassava, 27
Central African Federation, 33; Yoke of, 35-37
Central African Federation of Rhodesia and Nyasaland, 24
Central African Republic, 41
Chad, 41
Chile, 147
Chileshe, Alderman Safeli, H., 118
Chilombo, Anna, 88
Codmium, 20
Colonial administration, economic imbalance created by, 81-84; failure of, in the field of economic development, 82-83
Colonial Development and Welfare Acts, 58
Copper, 20
Copper industry, 98-99
Credit Organization of Zambia, 107, 168; failure of, 108
Cuba, 152

Deguefe, Taffara, 141
Development, plans, objectives of, 58-59; strategies, objectives of, 189-90

East African Community, 155
Eastern Block, 148
Eastern Province, 87
Economic Commission for Africa, 56, 176
Economic Commission for Africa, 56, 176
Economic development, central issues concerning, 178; definition of, 58; patterns of, 176-77; theory, 99
Economic development plans, development models and, 176-79
Economic diversification, 147
Emeralds, 20
Emergency Development Plan, 59, 60
Employment and Unemployment, solution to, 153-54
Ethiopia, 71

Federal Fiscal Commission, 37
Federation of Rhodesia and Nyasaland, 59, 109, 155
First National Development Plan, 59, 60, 71, 99, 128, 152, 177, 181; objectives of, 60-63
Food and Agriculture Organization of the United Nations 26, 27, 59

Forbes, P.W., 34
Foreign investment, role of, in Zambia's economy, 170-74
Friedman, Irving S., 146
Friedman, John F., 190

Galbraith, J.K., 127
Ghana, 93, 95; external debt of, 191
Gross Domestic Product, 21, 66, 153-54, 162, 174
Gross Fixed Capital, 128
Gross National Product, 17, 178; in Zambia, 148, 192

Halset, Walter, 153
Hanson, Harry, 141
Harvey, Charles, 112

Indigenization, 125; concept of, 116; reforms and, 116-20.
Indigenous Zambians, discrimination against, 79-80; in expatriate firms, 100; predominance of, 119
Industrial Development Act, 132
Industrialization, 155-57
Intensive Development Plan, 71
Intensive Development Zones, 66
International Bank for Reconstruction and Development, 44, 47, 59, 113
Israel, 152

Japan, 121, 127
Johnstone, Harry, 34
Jolly, Richard, 74

Kabwe, 83
Kalegwa, 129
Kasonde, Emmanuel G., 108, 169
Kashita, E.A., 103
Kaunda, Kenneth, 38
Kenya, 93, 95, 131; agriculture in, 194; industrial activities in, 189
Keynes, John Maynard, 122
Kitwe National Convention, 65
Korea, 121

Land development, issues of, 160-62
Lengalenga, Anna, 88

Lesotho, 41
Less-developing countries, developed countries and, 57
Livingstone, David, 18, 40, 67, 98
Livingstone Fiat Motor Assembly, impact of, on Zambian economy, 171-73
Lochner, 33
Lopez, Odoardo, 18
Lozi, 82
Luapula Provinces, 16; business pioneers of, 87; commercial activities in, 86
Lugard, Frederick, 82
Lukanga, 16
Lusaka, 83, 88

Maize, 27
Malawi, 41; growth rate of agriculture in, 194
Mali, 41
Malinki, Tanner, 89
Mansa, 88, 107
Marx, Karl, 177; theories in Third World African countries, 177
McNamara, Robert, 178
Menaco, 185
Metro Economic Reform Programme, 99, 118-19, 120, 122, 128, 133; contribution of in sharpening the Zambian economy, 109-16; directions to mining industry, 110-13; mining sector and, 109-10
Mill, John, 58
Millet, 27
Mimbula, 129
Mineral marketing, 114-16
Mining Affairs Appeal Tribunal, 130
Mining Development Corporation Limited, 109, 111-12, 177; objectives of, 129-30
Mining industry, 91, 94
Monazite, 20
Mongu, 67
Morocco, 148
Morris, C.T., 178
Mozambique, 43
Mukupo, Titus B., 89

Index

Mulungushi Economic Reform Programme, 99, 101, 119-20, 122, 133; activities of, 104; contribution of, in Zambian economy, 100-109; and Indeco, 102-105; and large scale financing, 107-109; small scale financing and, 105-107; Zambian economy before, 100-101
Mumba, Luka, 107
Musakanya, Flavia, 169

National Agricultural Marketing Board, 121, 137, 160, 166-67, 181
National Transport Corporation, 136, 167; subsidiaries of, 136
New Caledonia, 148
New International Economic Order, 57
Nickle, 20
Nigar, 41
Nigeria, 93, 95
Northern Rhodesia, 102
North Eastern Rhodesia, 34, 40
North West Rhodesia, 33, 40

Ohadike, Patrick, 28
Oppenheimer, Haroy, 119
Organization of African Unity, 187

Pan-African Telecommunication Network, 159
Parastatals, activities relating to transport and communications 136-37; agriculture sector and, 137-39; legal inadequacies of using for development, 141-44; legal and sectoral analysis of, 127-28; national economy and, 166-70
Pigafelta, Filipo, 18
Planning, social approach to, 190-91
Poland, 177
Political independence, integrating with development, 89-96
Prain, Ronald, 128
Pebisch, Raul, 56
Public Corporation, definition of, 124; conceptual framework of, 125-27; objectives of, 124; role of, in state economy, 125; Zambia's policy towards the establishment of, 127
Public enterprises, importance of, in economic growth, 180-85; public involvement in, 181; structural changes in, 182-83, 185
Pyrochiore, 21

Rhodes, Cecil John, 33, 35
Rice, 27
Roan Selection Trust, 91, 111, 112, 118, 128; mining operation of, 118-19; recognition to, 90
Rodney, Walter, 35, 57
Rostow, Walter, 56
Rousseau, Jean Jacques, 58
Rwanda, 41

Schumacher, E.F., 197
Second National Development Plan, 59, 64, 128, 152, 154, 177, 181; agriculture sector during, 69-70, 162; air transport system during, 69; economic growth during 66-69; objectives of, 63-64
Seidman, Ann, 96
Sesheke, 67
Sharpe, Alfred, 34
Simwinga, George, 112
Smith, Adam, 58
Smith, Ian, 42, 46
Somalia, 71
South Africa, 148
Southern Rhodesia, 27, 43
Sri Lanka, 152
State Finance and Development Corporation, 108, 134, 168, 177; subsidiaries of, 108

Taiwan, 127
Tanzania, 41, 93, 131; industrial activities in, 189
Tanzania Zambia Railway, 66, 137; IBRD refusal to finance, 47
Tarrys, E.W., 106
Temu, Peter, 161
Thermaat, J. Verloren van, 121
Third World, 187

Third World Countries, 115, 124, 190; development prospects for, 191; growth rate of agriculture in, 194; socialist rhetoric and, 177; theories of development plan in the plans of, 51-59
Thompson, Joseph, 34
Transitional Development Plan, 59, 60

Unemployment *see* Employment and unemployment
Unilateral Declaration of Independence, 62, 63, 74, 94, 136, 151, 177; a duel between Britain and Rhodesia, 43
United Kingdom, 42, 131
United Nations, 44, 56, 90, 156, 191
United Nations Conference on the Law of the Sea, 41-42
United Nations Conference on Trade and Development, 178
United Nations Economic Commission for Africa, 26, 51, 59, 187
United Nations Educational Scientific and Cultural Organization (UNESCO), 59
United Nations Independence Party, 92, 93, 100, 169
United Nations Industrial Development Organization, 187
United States of America, 127, 147-48
USSR, 177

Wallace, L.A., 34
Ware, 33
Wilson, Harold, 42
World Bank, 191

Yogoslavia, 152

Zaire, 41, 147-48, 155
Zambia, agricultural activities in, 26-28; business activities in, 88-89; colonial period of, 33-35; copper resources of, 147-48; development activities of, 146; development plans in, 59-65; economic growth and development bases for, 17-18; economic reform process of, 71; education facilities in, 152-53; exports and imports of, 192; external debt of, 191; food products of, 27; geographical profile of, 14, 16, 17; hotel and restaurant business at, 88; industrial development of, 131-34, 186-88; industrialization of, 189; landlocked, political origin of becoming, 40-42; mineral resources of, 18-21, 24, 26; population structure of, 28-32; pre-independence economic imbalances of, 81-85; pre-independence native business in, 85-89
Zambia Airways, 137, 185
Zambian economy, by the end of first decade of independence, 185-86; challenges to, 193-94; country's resources and, 147-48, 151-52; difficulties faced by; 148-51; economic indicators and, 17, ills suffered by, 35-36; improvement in, after independence, 94; issues related to integrating the, 92-93; mining dominance in, 128-31; mobilization of financial resources to promote, 134-36; new transit routes and, 47; parastatals and, 152; performance of, 191-93; political independence and, 90-91; remedies for challenges to, 195-96; UDI's effects on, 42-46.
Zimbabwe, 41.

Bookkeeper

Deacidification for Libraries and Archives

May 2008